Mastering Sustainable Agriculture: A Comprehensive Teacher's Guide to Eco-Friendly Farming Practices

Mastering Sustainable Agriculture: A Comprehensive Teacher's Guide to Eco-Friendly Farming Practices

Introduction

Chapter 1: Understanding Agriculture and Sustainability

Chapter 2: Soil Health and Conservation

Chapter 3: Water Management

Chapter 4: Crop Diversification

Chapter 5: Pest Management

Chapter 6: Livestock Management

Chapter 7: Technology in Sustainable Agriculture

Chapter 8: The Economics of Sustainable Agriculture

Chapter 9: Teaching Sustainable Agriculture

Conclusion

Introduction

In a rapidly changing world, the importance of sustainable agriculture cannot be overstated. As the global population surges towards 10 billion by 2050, the demand for food will also significantly increase. Conventional agricultural practices, although effective in the short term, often degrade the environment by eroding soils, depleting water sources, and contributing to climate change through excessive greenhouse gas emissions. These impacts are not just environmental; they are also social and economic. Communities around the world face threats to their livelihoods due to unsustainable agricultural practices that degrade the resources upon which they depend. Sustainable agriculture provides a viable alternative that aims to balance the need for food production with the preservation of environmental integrity. By adopting sustainable practices, farmers can increase yields and improve soil health, all while reducing their carbon footprint. These techniques often go hand-in-hand with social benefits like fair wages and community development. Therefore, adopting sustainable agriculture is not just an environmental imperative but also a necessity for social and economic resilience.

Target Audience

This guide is designed for educators who are looking to incorporate the principles of sustainable agriculture into their teaching curricula. The content is tailored for high school and college-level courses in agriculture, environmental science, and related subjects. In addition to educators, this guide may also be useful for agricultural extension officers, NGO workers, and policymakers interested in sustainable development. Furthermore, it can serve as a comprehensive resource for students eager to learn about the subject and for farmers who wish to adopt more sustainable practices in their operations. The guide provides in-depth information, case studies, and lesson plans that can be adapted to various educational settings and levels.

How to Use This Guide

Navigating through this guide is designed to be straightforward and user-friendly. Each chapter focuses on a specific aspect of sustainable agriculture, starting with foundational knowledge and gradually moving towards more advanced concepts and practical applications. Each section within a chapter is designed to be self-contained but also complementary to the overall theme. Case studies are included to offer real-world examples of how sustainable agriculture practices have been successfully implemented. These are excellent touchpoints for class discussions or individual research projects. Additionally, the guide includes various lesson plans and resources that educators can incorporate directly into their curricula. These materials are designed to be flexible, allowing you to pick and choose according to your needs and the specific outcomes you wish to achieve. The concluding chapter sums up key topics and offers additional resources for further study and community involvement. Whether you're an educator planning a semester-long course or a farmer seeking information on specific sustainable practices, this guide is a comprehensive resource designed to meet a variety of needs.

By adhering to the structure and guidelines laid out in this introduction, you'll be well-equipped to explore the crucial topic of sustainable agriculture in a meaningful way.

Chapter 1: Understanding Agriculture and Sustainability

As we venture into the 21st century, the significance of agriculture in our daily lives continues to grow. However, its environmental and social footprint is increasingly under scrutiny. This chapter aims to demystify the complexities surrounding agriculture and sustainability, laying a robust foundation for the rest of the guide. The concept of sustainability is often reduced to buzzwords and platitudes, but it's crucial to understand its full scope and applicability, especially in agriculture. This is not just about eco-friendly farming; it's about a holistic approach that considers economic viability, social equity, and long-term environmental health. The chapter will delve into the essential facets of agriculture, introduce the concept of sustainability, and compare traditional with sustainable agricultural practices. We will also present real-world case studies showcasing sustainable practices in action. By the end of this chapter, you will gain a solid understanding of what sustainable agriculture means and why it is crucial for the future of our planet and society. The goal is to arm educators, students, and practitioners with the knowledge needed to make informed decisions and advocate for sustainable practices in agricultural settings.

What is Agriculture?

Agriculture is the practice of cultivating land, raising crops, and managing livestock with the primary goal of producing food, fiber, medicinal plants, and other products that sustain and enhance human life. It is one of the oldest and most essential human activities, evolving over thousands of years to meet the needs of growing populations and changing societies. Agriculture has taken many forms throughout history, from small-scale subsistence farming to the industrial agricultural systems that dominate today's food production landscape.

The scope of agriculture extends far beyond the traditional notion of farming to include activities such as soil preparation, seed planting, crop management, harvesting, storage, and distribution. It also involves auxiliary sectors like research, development, and technology deployment aimed at improving crop yields and livestock health. The advent of biotechnology, automated machinery, and data analytics has transformed agriculture into a complex, technology-driven enterprise that requires specialized knowledge and skills.

Moreover, agriculture is deeply intertwined with other sectors of society, including economics, social structures, and politics. The food we eat, the clothes we wear, the medicines we consume, and even the fuel in our vehicles are products of agricultural systems. Agriculture is not just an economic activity but a cultural and social one as well, affecting everything from local traditions to global markets. For example, the cotton produced in one country may be processed in another and turned into garments in yet another country, all the while governed by international trade policies and regulations.

However, as the global population continues to grow, and the demand for food and other agricultural products increases, the sector faces unprecedented challenges. Environmental degradation, climate change, and dwindling natural resources pose threats to the long-term sustainability of agricultural systems worldwide. The current agricultural practices often rely heavily on fossil fuels, synthetic fertilizers, and pesticides, leading to soil degradation, water pollution, and loss of biodiversity.

In summary, agriculture is a multifaceted and evolving field that is central to human survival and societal development. While it has successfully met the food and material needs of billions of people, the methods by which these successes have been achieved are now posing challenges that need to be addressed urgently. Understanding the breadth and depth of agriculture is the first step towards creating more sustainable systems that can feed the world while also preserving the planet's health.

The Concept of Sustainability

Sustainability is a multidimensional concept that aims to meet the needs of the present without compromising the ability of future generations to meet their own needs. Though it initially emerged in the context of natural resource management and environmental conservation, the concept has evolved to encompass economic and social dimensions as well. At its core, sustainability is about balance—between human needs and environmental limits, between economic development and social equity, between short-term gains and long-term consequences.

In the agricultural context, sustainability means implementing farming practices that are ecologically sound, economically viable, and socially just. This includes minimizing environmental impact, reducing dependency on non-renewable resources, enhancing soil fertility, and promoting fair labor practices. But sustainability is not a one-size-fits-all approach; it can differ depending on geographic, economic, and cultural factors. For example, sustainable agriculture in a developed country might involve the use of high-tech solutions like precision farming, while in a developing country, it might be more focused on traditional practices that have low environmental impact.

One key aspect of sustainability is the focus on long-term goals over short-term gains. This involves recognizing the interconnectedness of ecosystems and understanding that actions taken today have consequences that can last for generations. It also necessitates a shift from a linear model of production-consumption-waste to a more circular economy where resources are reused, recycled, or regenerated.

Another crucial aspect is stakeholder involvement. True sustainability cannot be achieved in isolation; it requires the active participation of all stakeholders—from farmers and consumers to policymakers and researchers. Transparent governance, community engagement, and knowledge-sharing are all integral to achieving a more sustainable future for agriculture.

While the concept of sustainability has gained mainstream attention, it is often misinterpreted or oversimplified. For many, it's reduced to mere buzzwords or marketing strategies. However, the essence of sustainability is a comprehensive understanding that integrates environmental, social, and economic factors. It acknowledges that complex problems require multifaceted solutions and that every action has a ripple effect, impacting people and the planet. As educators, it's crucial to impart this nuanced understanding of sustainability to students, who are the future stewards of our planet, to ensure that they are equipped with the knowledge and tools to make more sustainable choices in agriculture and beyond.

Importance of Sustainable Agriculture

Sustainable agriculture is vital for numerous reasons, primarily as it addresses the trio of environmental, economic, and social sustainability while contributing to food security. As the global population approaches 8 billion and shows no signs of slowing down, the pressure on agricultural systems to provide adequate food is immense. Traditional agriculture often relies on practices that are environmentally detrimental, such as deforestation, excessive water usage, and heavy application of chemical fertilizers and pesticides. These approaches are not only unsustainable in the long run but also pose immediate threats like soil erosion, water pollution, and loss of biodiversity.

From an environmental standpoint, sustainable agriculture aims to preserve natural resources for future generations. It promotes the use of practices like crop rotation, mixed farming, and organic farming, which maintain or improve soil fertility, reduce water consumption, and minimize pollution. These practices also contribute to climate change mitigation by reducing greenhouse gas emissions associated with agriculture.

Economically, sustainable agriculture offers multiple benefits. It often requires lower inputs of expensive synthetic chemicals and promotes the use of locally available resources, thereby reducing costs for farmers. In the long run, sustainable agriculture can increase yields by improving soil fertility and water efficiency. It can also open up new market opportunities for farmers, as there is a growing consumer demand for sustainably produced foods. Moreover, these practices make farming systems more resilient to environmental stresses, thereby reducing the vulnerability of rural communities to climate-related disasters.

On the social front, sustainable agriculture supports fair labor practices and community well-being. It emphasizes the need for farmers to earn a living wage and promotes gender equality by involving both men and women in farming activities. Moreover, it places a focus on community engagement and shared decision-making, empowering local communities to have a say in the management of their resources.

Lastly, sustainable agriculture supports animal welfare by promoting ethical treatment and proper living conditions for livestock, leading to healthier animals and, in turn, better quality meat and dairy products.

Therefore, the importance of sustainable agriculture extends far beyond merely producing food. It plays a critical role in safeguarding our planet, promoting economic viability, and ensuring social justice. Implementing sustainable agriculture is not just a choice but a necessity for the well-being of future generations. It is a holistic approach that, when implemented thoughtfully, can address some of the most pressing challenges of our time.

Traditional vs. Sustainable Agriculture

Traditional and sustainable agriculture are two distinct approaches to farming that have their own merits and drawbacks. Understanding the key differences between the two can provide valuable insights into the future of agriculture and its potential to meet global food needs in an environmentally friendly and socially responsible manner.

Traditional agriculture, often characterized by large-scale, monoculture farming, has been incredibly effective at increasing food production to meet the demands of a growing global population. This approach typically employs synthetic fertilizers, pesticides, and genetically modified organisms (GMOs) to achieve high yields. While successful in producing abundant food supplies, this form of agriculture is increasingly criticized for its environmental impacts, including soil degradation, water pollution, and greenhouse gas emissions. In addition, it often relies on economies of scale, which can displace small-scale farmers and contribute to social inequalities.

On the other hand, sustainable agriculture focuses on long-term environmental stewardship and social responsibility. It seeks to minimize negative environmental impacts by using techniques like crop rotation, intercropping, and organic farming. These methods help maintain soil fertility, reduce the need for chemical inputs, and increase resilience against pests and diseases. In addition, sustainable agriculture aims for economic viability, ensuring that farmers can make a living without exploiting natural resources. It often incorporates ethical labor practices and community engagement, fostering social cohesion and mutual respect among stakeholders.

One point of contention between the two approaches is productivity. Critics of sustainable agriculture argue that it cannot meet the world's growing food demands because it generally produces lower yields compared to traditional methods. However, proponents counter that sustainable practices are more resilient to climate change and other environmental factors, making them more reliable in the long run. Moreover, sustainable agriculture looks beyond yield as the only measure of success and takes into account other factors like environmental health, social equity, and long-term viability.

Another noteworthy difference is the treatment of livestock. Traditional agriculture often employs factory farming techniques, which raise ethical and health concerns, while sustainable practices focus on free-range, organic, or pasture-raised methods that are better for both the animals and consumers.

In summary, traditional and sustainable agriculture serve different objectives and have varying impacts on the environment, economy, and society. As the limitations of traditional agriculture become more evident, there's an increasing shift toward more sustainable practices. The challenge lies in balancing the strengths and weaknesses of each approach to create a more resilient and equitable global food system.

Case Studies: Sustainable Practices in Action

1. Polyface Farms, Virginia, USA: Owned by Joel Salatin, Polyface Farms is a prime example of regenerative agriculture, an advanced form of sustainable farming. The farm employs

rotational grazing and symbiotic animal and crop systems. Cows graze the land first, followed by chickens that feed on the cow droppings, thereby breaking down the manure into nutrient-rich soil. This practice not only enhances soil fertility but also reduces the need for chemical fertilizers. Polyface Farms has become a model for sustainable agriculture, offering both educational programs and farm tours.

2. Sikkim, India – Organic State: In 2016, the Indian state of Sikkim became the first fully organic state in the world. All synthetic pesticides and fertilizers have been banned, and the state has adopted sustainable practices like crop rotation and composting. As a result, the soil's health has improved, and the farmers are getting better prices for their organic produce. This shift has not only benefitted the environment but also the social and economic well-being of the local farmers.

3. Fair Trade Coffee in Colombia: Small coffee growers in Colombia have embraced Fair Trade practices that emphasize sustainable agriculture and social responsibility. These farmers employ agroforestry methods, planting coffee under the shade of native trees, which helps prevent soil erosion and promotes biodiversity. This approach has led to better quality coffee and improved incomes for the farmers, demonstrating that sustainability and profitability can go hand-in-hand.

4. Community-Supported Agriculture (CSA) in the UK: CSA programs have gained popularity across the United Kingdom as a sustainable alternative to large-scale farming. Consumers pay farmers in advance for a season's worth of produce, which reduces the financial risks for the farmer and encourages sustainable practices. These direct relationships between producers and consumers foster a sense of community and offer a transparent food supply chain.

5. Kibbutz Lotan, Israel: Located in Israel's arid southern region, Kibbutz Lotan practices permaculture and sustainable

agriculture, emphasizing conservation and efficient use of water through drip irrigation and other innovative techniques. They also run educational programs to teach sustainable practices to people from around the world.

These case studies showcase the broad range of sustainable agriculture practices that are not only environmentally friendly but also economically viable and socially equitable. They serve as inspirational models for farmers, policymakers, and consumers alike, demonstrating that sustainable agriculture can indeed be the way forward for a more resilient and ethical food system.

Summary

Chapter 1 aimed to provide a comprehensive understanding of agriculture and sustainability. Starting with the definition of agriculture, we delved into its various aspects, from food production to its role in shaping human civilizations. Subsequently, the concept of sustainability was discussed as a critical need for both the environment and society. The importance of sustainable agriculture was emphasized, elucidating how it can provide a solution to many of the issues plaguing traditional agriculture, such as soil degradation, water scarcity, and loss of biodiversity.

Comparing traditional and sustainable agriculture, we highlighted the significant differences between the two, from the use of chemical inputs to environmental impact. Real-life case studies were presented to give practical examples of sustainable practices in action, ranging from small-scale local farms to large-scale initiatives like Sikkim's transformation into an organic state.

This chapter serves as a foundation for understanding the why and how of sustainable agriculture. The information presented here is essential for educators, policymakers, and individuals interested in making a meaningful impact in the realm of agriculture. The examples underscore the viability and necessity of adopting sustainable practices for a more resilient and ethical food system.

Chapter 2: Soil Health and Conservation

Chapter 2 delves into one of the most crucial components of sustainable agriculture: soil health and conservation. Soil is more than just a medium for plant growth; it is a living, dynamic system that provides essential nutrients and serves as a foundation for all terrestrial life. Unfortunately, conventional farming practices like excessive tilling, monocropping, and the overuse of synthetic fertilizers and pesticides have led to widespread soil degradation. The loss of fertile soil is not only an agricultural issue but also a significant environmental and social concern.

In this chapter, we will explore the various aspects of soil health, from its composition and importance to techniques aimed at its conservation. Educators will find information on how to teach students the science behind soil conservation and its role in sustainable agriculture. We will also look at the relationship between soil health and other ecological factors like water quality and biodiversity. Real-world case studies will offer practical examples of how soil conservation techniques are being successfully implemented across different scales and geographies.

By understanding and respecting the role of soil in sustainable agriculture, we can move towards more responsible and sustainable farming practices.

The Importance of Soil Health

Soil health is an indispensable aspect of a well-functioning agricultural ecosystem. Healthy soil doesn't just facilitate the growth of crops; it acts as a holistic system that influences water quality, aids in climate change mitigation, and supports a complex web of life. In a sustainable agricultural context, soil health is crucial for multiple reasons.

Firstly, soil rich in organic matter and microorganisms provides essential nutrients to plants, negating the need for synthetic fertilizers. These nutrients not only ensure optimal plant growth but also contribute to the nutritional content of the produce. A well-balanced soil ecosystem can improve crop yield and quality, leading to more nutritious food for consumers.

Secondly, healthy soil plays a vital role in water management. It acts as a natural filter, purifying water as it passes through the soil layers. Moreover, good soil structure enhances water retention, reducing the need for frequent irrigation and mitigating the impact of drought conditions. Soil health is intrinsically linked to water conservation, an increasingly crucial issue given the rising incidences of water scarcity globally.

Thirdly, healthy soil is a natural carbon sink. Organic matter within the soil captures and stores carbon dioxide, thus aiding in climate change mitigation. According to the Food and Agriculture Organization (FAO), sustainable soil management can contribute to storing up to 10% of global carbon dioxide emissions.

Fourth, soil health also relates to biodiversity. A healthy soil ecosystem is teeming with life, from microscopic bacteria and fungi to larger organisms like earthworms and beetles. These organisms play essential roles in nutrient cycling, disease suppression, and organic matter decomposition. They are integral to the soil food web, and their presence is often an indicator of soil vitality.

Lastly, understanding and maintaining soil health is also an economic imperative. Healthy soils can reduce input costs for farmers by decreasing the need for irrigation and synthetic fertilizers. Over time, sustainable soil management can result in more resilient agricultural systems that are less susceptible to market fluctuations and extreme weather events.

In summary, the importance of soil health in sustainable agriculture cannot be overstated. It's a linchpin for environmental, social, and economic sustainability, connecting multiple facets of agriculture into a cohesive, functioning system. Teachers imparting this knowledge to the next generation are not merely teaching science but are also equipping them with the understanding required to create a more sustainable future.

Techniques for Soil Conservation

Soil conservation is not a one-size-fits-all approach but rather a collection of practices tailored to specific needs and local conditions. The following are some effective techniques that can be employed to improve and maintain soil health. These methods are adaptable and can be introduced in classroom settings to instill the importance of soil conservation in students:

- Crop Rotation: Changing the types of crops grown in a specific field over time can help prevent soil depletion. Different crops have different nutrient needs, and rotation helps in balancing the soil's nutrient profile.
- Cover Crops: Growing cover crops like legumes and grasses can prevent soil erosion and improve soil structure. These plants also fix nitrogen in the soil, reducing the need for synthetic fertilizers.

- Reduced or No-Till Farming: Traditional tilling disrupts the soil structure and leads to erosion. Reduced or no-till methods maintain the soil's integrity and enhance its organic matter content.
- Contour Farming and Terracing: For sloping lands, contour farming can help in reducing soil erosion. Terracing can also be employed to create flat platforms, making it easier to cultivate and lessening soil run-off.
- Riparian Buffers: Planting vegetation like trees and shrubs along water bodies can act as a buffer, preventing soil from entering waterways, which in turn protects both soil and water quality.
- Use of Organic Matter: Adding compost or other organic matter enhances soil structure, improves water retention, and provides essential nutrients.
- Integrated Pest Management (IPM): Using biological controls and natural pesticides reduces the need for chemical pesticides, which can degrade soil health.
- Agroforestry: Incorporating trees into crop systems can help in soil conservation by reducing wind and water erosion and improving soil organic matter.
- Mulching: A layer of organic or inorganic material can protect the soil from the impact of rain, reducing erosion and improving water retention.
- Soil Testing: Regular soil tests can provide valuable information about its nutrient levels, pH, and other properties, guiding appropriate and sustainable soil management practices.

Implementing these techniques doesn't require massive changes to existing farming practices but does demand a shift in perspective. By integrating these soil conservation methods into the educational curriculum, teachers can prepare students to be stewards of the land, understanding the need for and methods of maintaining soil health for future generations.

Organic Matter and Soil

Organic matter is the lifeblood of fertile, healthy soil. Composed of decomposed plant and animal residues, microbial biomass, and root exudates, organic matter plays a critical role in enhancing soil properties and supporting sustainable agricultural practices. Teachers aiming to incorporate lessons on sustainable agriculture should place a strong emphasis on the importance of organic matter in soil health.

Firstly, organic matter serves as a reservoir of nutrients. As organic material decomposes, essential nutrients like nitrogen, phosphorus, and potassium are released into the soil. These nutrients are readily available for plant uptake, reducing the need for chemical fertilizers that often lead to soil degradation and environmental pollution.

Secondly, organic matter significantly improves soil structure. Well-structured soil has better water infiltration, increased water-holding capacity, and reduced runoff and erosion. This is crucial in regions prone to drought or heavy rainfall, as good soil structure can mitigate the effects of these extreme weather events on crop yields.

Thirdly, organic matter enhances soil's water-holding capacity. A higher amount of organic material allows soil to retain more water, which can be especially beneficial in dry regions or during periods of low rainfall. This not only conserves water but also ensures a consistent supply to the plants, thereby sustaining yields.

Moreover, organic matter is central to the soil food web, hosting a range of microorganisms like bacteria, fungi, and earthworms. These organisms break down organic material, recycle nutrients, and contribute to soil structure, creating a balanced ecosystem that supports plant growth and resilience against diseases.

Additionally, soils rich in organic matter have higher cation-exchange capacity (CEC). Higher CEC means the soil can hold onto essential nutrients longer, making them available for plants and reducing nutrient leaching into waterways.

Teachers can use practical classroom activities to demonstrate the role of organic matter in soil health. Simple experiments showing water retention in soils with different organic matter levels or the rate of decomposition of various organic materials can be effective teaching tools.

In summary, organic matter is indispensable for the maintenance of soil health and, by extension, sustainable agriculture. Whether in classroom instruction or practical applications, understanding the value of organic matter lays the groundwork for more conscientious and effective agricultural practices.

The Role of Microorganisms

Microorganisms are the unseen powerhouses driving the health and fertility of soil. These microscopic entities, including bacteria, fungi, protozoa, and nematodes, play multifaceted roles in soil ecology and are central to the concept of sustainable agriculture. Teachers introducing sustainable farming methods to their students must include the vital roles that soil microorganisms play.

First and foremost, soil microorganisms are critical to nutrient cycling. They decompose organic material, breaking it down into essential nutrients that plants can absorb. For example, nitrogen-fixing bacteria convert atmospheric nitrogen into forms usable by plants, reducing the need for synthetic nitrogen fertilizers. Similarly, mycorrhizal fungi form symbiotic relationships with plant roots to improve phosphorus uptake.

Microorganisms also contribute to soil structure. Through their metabolic activities and the substances they produce, they create stable soil aggregates, which improve water infiltration and retention. This is especially important for sustainable practices that aim to conserve water and reduce erosion.

Furthermore, soil microorganisms play a role in suppressing plant diseases. They produce natural antibiotics that inhibit the growth of pathogenic microbes, thereby reducing the need for chemical pesticides. This is a significant point for sustainable agriculture, as reducing chemical input is not only cost-effective but also less harmful to the environment.

Microbial communities can be influenced by agricultural practices. For example, tilling can disrupt these communities, while organic farming tends to support a more diverse microbial population. Understanding these relationships is crucial for making sustainable choices in agricultural practices.

In the educational context, teachers can utilize experiments to visualize the presence of microorganisms in soil samples using basic microscopy techniques. Students can also set up experiments to understand the impact of different agricultural practices, like tilling or organic farming, on microbial communities.

In summary, the role of microorganisms in soil health is a subject of paramount importance in sustainable agriculture. Through nutrient cycling, soil structuring, and disease suppression, they contribute significantly to the sustainability and productivity of agricultural systems. For educators, highlighting the essential role of soil microorganisms offers a tangible, engaging way to convey the importance of soil health in sustainable farming.

Case Studies: Soil Conservation

Understanding the importance of soil health is best emphasized through real-world examples. Let's delve into two case studies that illuminate how soil conservation techniques have been successfully implemented, providing tangible benefits for both farmers and the environment.

The Loess Plateau, China

Once considered an ecological disaster due to severe erosion, the Loess Plateau underwent a massive transformation through a focused soil conservation effort. Farmers shifted from traditional monocropping to terracing and agroforestry. By planting trees and diversifying crops, the project was able to significantly reduce soil erosion. The most notable outcome was increased agricultural productivity, which subsequently led to better living standards for the local population. Educators can teach students about large-scale transformations, focusing on how collective efforts and governmental support can make a significant impact.

Rodale Institute, United States

This research institute has long been a pioneer in organic farming methods, with a particular emphasis on soil health. Through long-term studies, they demonstrated that organic farming systems, rich in compost and organic matter, were more resilient to climate change due to better water retention and reduced soil erosion. The institute's "Farming Systems Trial" is the longest-running comparative study of organic and conventional agriculture in America. The findings of this study have shaped federal policies and inspired a new generation of farmers.

These case studies can serve as robust educational tools. Teachers can create activities that require students to analyze the methods employed, discuss the challenges faced, and consider the broader implications for sustainable agriculture. Hands-on activities could include model-building of terraced farming or research projects on the benefits of organic farming for soil health.

In summary, real-life case studies provide valuable educational material for understanding the practical aspects of soil conservation. They offer inspiration and concrete examples that sustainable agricultural practices can yield economic, environmental, and societal benefits. These cases can serve as the basis for in-depth discussion, further research, and practical experiments in educational settings.

Summary

This chapter on Soil Health and Conservation has aimed to provide a comprehensive understanding of the multifaceted aspects of soil as an essential component in sustainable agriculture. Starting with the vital role soil plays in agricultural ecosystems, the chapter discussed the techniques for soil conservation, the importance of organic matter, and the often-overlooked role of microorganisms in soil health.

Through real-world case studies, we observed practical applications of soil conservation methods, like terracing and organic farming, that not only safeguard the environment but also improve agricultural productivity. These examples serve as excellent educational resources for teachers aiming to bring real-world context into the classroom.

As we conclude, it's crucial to emphasize that soil health is not a stand-alone topic but an integral part of the broader sustainability conversation. Soil conservation is interlinked with water management, crop diversification, and pest management, among other topics, each contributing to the larger aim of sustainable agriculture. The health of our soil directly impacts the food we grow and the planet we live on.

Therefore, understanding and teaching about soil health are not just academic pursuits but critical for our collective future.

Chapter 3: Water Management

Welcome to Chapter 3, focusing on Water Management, an indispensable component of sustainable agriculture. As the world grapples with increasing water scarcity and climate unpredictability, understanding how to manage water resources efficiently has never been more crucial. For educators, this topic provides a fertile ground for discussions on resource conservation, environmental science, and even social ethics.

In this chapter, we will delve deep into the significance of water in agricultural settings, from nurturing plants to maintaining livestock and soil health. We will explore innovative irrigation methods that aim for optimal water use and examine strategies for water recycling in agriculture, a necessity for sustainability. Furthermore, the chapter will include case studies that showcase successful water management practices, providing real-world examples to make the teaching experience more interactive and enlightening.

By the end of this chapter, teachers will be equipped with the knowledge and resources to effectively teach about water management in the context of sustainable agriculture. It's not just about learning how to save water but understanding how intelligent water use contributes to a more sustainable and equitable world.

Importance of Water in Agriculture

Water is often referred to as the lifeblood of agriculture. It serves multiple functions, from germinating seeds to facilitating nutrient uptake, supporting plant growth, and enabling various agricultural activities. It is an essential resource without which sustainable farming practices would be impractical, if not impossible. This section aims to delve into the critical role that water plays in agriculture and why its efficient management is pivotal for sustainability.

First and foremost, water is necessary for crop growth and survival. The water cycle, including processes like evaporation, transpiration, and precipitation, creates the environmental conditions that help plants grow. Water is also instrumental in transporting essential nutrients from the soil to the plant cells. When water becomes scarce, the nutrient uptake is hampered, leading to weakened plants and reduced yields.

Water also plays a vital role in livestock management. It is needed not only for drinking but also for other uses like sanitation, food production, and temperature control. The amount of water required can be immense, especially when you consider that water consumption in livestock farming can account for almost 8% of global water usage.

Another area where water is critical is soil management. Water aids in breaking down organic matter and contributes to the soil structure. Well-structured soil has better water-holding capacity, which minimizes runoff and soil erosion, thereby contributing to a sustainable farming ecosystem.

Irrigation systems, a cornerstone in modern agriculture, are reliant on water availability. However, traditional methods can be inefficient, wasting much of the water applied. Therefore, understanding the importance of water also includes embracing more efficient irrigation techniques, which we will discuss later in this chapter.

Moreover, water is key in various agricultural processes like fertilization and pesticide application. Water-based solutions are commonly used for distributing these essential additives, which means that water quality can directly impact the effectiveness of these agricultural inputs. Poor water quality can lead to reduced efficiency in fertilizer and pesticide use, raising the cost of agricultural activities and reducing their sustainability.

It's not an exaggeration to say that every aspect of agriculture is intertwined with water use. Its importance transcends beyond just being a component of the growth medium to being an enabler of almost every process in farming. Therefore, the efficient and responsible management of water resources is not only an environmental necessity but also an economic imperative for sustainable agriculture.

For educators, understanding the importance of water in agriculture allows for the creation of compelling lesson plans that inspire students to think critically about resource management. It sets the stage for discussing innovative solutions and emphasizes why water management should be at the heart of any conversation about sustainable agriculture.

Efficient Irrigation Methods

Irrigation is a crucial part of agriculture, especially in regions where natural water supply is insufficient for optimal crop growth. However, traditional irrigation methods often lead to water waste, energy inefficiency, and even soil degradation. In the context of sustainable agriculture, adopting efficient irrigation methods is imperative. This section outlines some of the most effective irrigation techniques that can be incorporated into sustainable farming practices:

- Drip Irrigation: One of the most water-efficient methods, drip irrigation, delivers water directly to the plant's root zone. This minimizes water loss due to evaporation and runoff, and it allows for precise application of water and nutrients. Drip systems can be automated and easily managed, making them ideal for both small-scale and commercial farming.

- Sprinkler Systems: Unlike traditional flood irrigation, modern sprinkler systems can be customized to release specific amounts of water. They can also be set up to operate at times when water loss due to evaporation is minimal. Newer systems even have sensors to determine soil moisture levels, ensuring water is only applied when needed.
- Subsurface Irrigation: This method involves burying a network of tubing below the soil surface. Water is supplied directly to the root zone, similar to drip irrigation, but without any exposure to the surface, reducing evaporation loss. This is particularly useful in arid regions.
- Center Pivot Irrigation: Primarily used in large-scale farming, center pivot systems consist of rotating sprinkler arms that provide even water coverage. These systems can be equipped with variable rate irrigation technology to apply water more precisely, reducing wastage.
- Rainwater Harvesting: While not an irrigation method per se, rainwater harvesting can significantly supplement irrigation needs. Captured rainwater is stored and used for irrigation, reducing the reliance on other, often less sustainable, water sources.
- Hydroponics and Aquaponics: These soil-less farming methods use nutrient-rich water to grow plants, often recycling the same water multiple times. While the initial setup can be expensive, they are incredibly water-efficient in the long run.

Adopting these efficient irrigation methods can result in several benefits:

- Water Conservation: Obviously, the most immediate impact is a significant reduction in water usage, critical in areas facing water scarcity.

- Energy Savings: Efficient systems often require less energy to operate. For instance, drip systems work effectively at lower pressures compared to sprinkler systems, resulting in lower energy consumption.
- Increased Yield: Better water management generally leads to healthier plants and potentially higher yields, making the investment in efficient systems economically viable.
- Reduced Soil Erosion: By avoiding water excess, techniques like drip and subsurface irrigation can help preserve soil structure and reduce erosion.

For educators teaching sustainable agriculture, introducing these methods in the curriculum offers practical insights into modern farming challenges and solutions. Students should not only learn the science behind these methods but also the economics, understanding that investing in efficient irrigation is both an environmental and economic win-win.

Water Recycling in Agriculture

Water recycling is an integral component of sustainable agriculture, aimed at maximizing the utility of water resources while minimizing waste. Through various technologies and techniques, recycled water can be used for irrigation, livestock, cleaning, and even recharging groundwater. Given the increasing challenges of water scarcity and climate change, integrating water recycling into agriculture has never been more critical:

- Greywater Systems: One of the most straightforward methods for water recycling is the use of greywater—water that has been used in sinks, showers, and washing machines but not contaminated by human waste. Greywater can be directly used for irrigating non-edible plants and can also be treated for use on edible crops. These systems are especially beneficial for small-scale farms and urban agricultural settings.

- Constructed Wetlands: These are engineered systems that mimic natural wetlands. They serve as biofilters that treat wastewater through a combination of mechanical and biological processes, including sedimentation, filtration, and microbial activity. The treated water can then be used for irrigation.
- Nutrient Recovery: In livestock farming, manure and wastewater are often considered waste products. However, they can be sources of essential nutrients for plants. Technologies like anaerobic digesters can treat these waste streams to not only recover nutrients but also to produce biogas, a renewable source of energy.
- Aquaponics and Hydroponics: In these systems, water is continuously circulated between fish tanks and plant beds. Fish waste provides nutrients for the plants, and the plants help clean the water, which is then returned to the fish tanks. This creates a closed-loop, nearly zero-waste system.
- Irrigation Runoff Capture: In traditional irrigation systems, excess water that runs off can be captured, treated, and reused. Newer smart irrigation systems even include sensors that measure the amount of runoff and adjust water flow accordingly.

Benefits of Water Recycling

The benefits of water recycling include:

- Sustainability: The foremost advantage is the sustainability of water resources. It helps in lowering the dependency on freshwater sources, which are increasingly becoming scarce.
- Cost-Effectiveness: Initial set-up costs are outweighed by the long-term savings on water bills. For large agricultural operations, this can translate into significant financial gains.

- Reduced Pollution: Properly managed water recycling systems can reduce the discharge of pollutants into water bodies, thereby helping to protect aquatic ecosystems.
- Enhanced Productivity: With a more reliable water supply, farming activities can continue unhindered, potentially leading to increased productivity and yields.
- Climate Resilience: Recycling water can be a part of a broader strategy to make agricultural systems more resilient to climate change, particularly in regions that are increasingly experiencing drought conditions.

Water recycling not only serves to conserve valuable water resources but also embodies the very principles of sustainable agriculture—efficiency, circularity, and sustainability. It's a topic that warrants serious consideration in any curriculum focused on modern, sustainable agricultural practices. By teaching the next generation of farmers and agriculturists about these practices, we contribute to a more sustainable and secure future.

Case Studies: Water Management Practices

Case Study 1: Drip Irrigation in Israel

Israel has long been a pioneer in water management, especially given its arid climate and limited freshwater resources. One of the most effective water-saving technologies they've implemented on a large scale is drip irrigation. By delivering water directly to the roots of plants, this technology has been able to cut water usage by up to 50% compared to traditional methods. Besides conserving water, drip irrigation also minimizes waterborne diseases and weeds. Today, this technology is not only used across Israel but is also being adopted globally, especially in regions facing severe water scarcity.

Case Study 2: California's Greywater Systems

California has been facing drought conditions for several years, which has put a lot of pressure on its agricultural sector. To adapt, some farmers and households have started using greywater recycling systems. These systems collect water from household sinks, bathtubs, and washing machines and use it for landscape irrigation. This reduces the need for freshwater and lightens the load on sewage treatment facilities. Moreover, this practice has been legalized and encouraged by local governments, making it a mainstream water-saving solution.

Case Study 3: Rainwater Harvesting in India

Rainwater harvesting is an ancient practice that has received renewed attention in water-stressed regions of India. Several farmers have implemented systems that collect and store rainwater during the monsoon season for use during the dry months. This practice not only provides a sustainable water source but also helps in recharging local aquifers. The government has taken steps to subsidize the installation of rainwater harvesting systems, making it an economically viable option for small-scale farmers.

Case Study 4: Nutrient Recovery in the Netherlands

Dutch farmers have been at the forefront of sustainable agriculture, and their practices in nutrient recovery are no exception. Anaerobic digestion facilities treat livestock manure to extract not just biogas for energy but also valuable nutrients like nitrogen and phosphorus, which are then used as natural fertilizers. This method serves a dual purpose: it recycles waste material and lessens the dependency on chemical fertilizers, thereby reducing both waste and pollution.

Case Study 5: Smart Irrigation in Australia

In Australia, where water resources are exceptionally scarce, smart irrigation technologies are gaining traction. These systems use sensors and data analytics to measure soil moisture levels, weather conditions, and plant needs to optimize irrigation schedules. Some of these technologies even use AI to make real-time adjustments, ensuring not a drop of water is wasted.

Each of these case studies showcases how innovative water management practices can lead to more sustainable agriculture. They demonstrate the adaptability and resilience of farmers worldwide in tackling one of the most pressing issues of our time: water scarcity. By learning from these real-world examples, we can equip ourselves with the knowledge and tools needed to make more responsible choices in water management.

Summary

Water management stands as a cornerstone in the advancement of sustainable agriculture. As we've explored in this chapter, the importance of water in agriculture cannot be overstated, both for crop and livestock farming. Efficient water use not only conserves this precious resource but also leads to higher yields and more resilient farming systems. The methods discussed—from drip irrigation to greywater systems—provide actionable solutions for farmers facing various challenges related to water scarcity.

The real-world case studies illustrate the remarkable strides that have been made in this field. Countries and communities across the globe, from Israel to India, have adopted innovative approaches to make the most out of their water resources. These examples also underscore the role of technology and government policies in facilitating sustainable water management practices. Technologies like smart irrigation systems and nutrient recovery techniques offer a glimpse into the future of farming, one where agriculture and sustainability go hand in hand.

As we move forward, it is essential to disseminate these best practices and technologies more widely. This will not only help individual farmers but also contribute to the broader goals of environmental conservation and food security.

Chapter 4: Crop Diversification

As educators navigating the ever-evolving landscape of agriculture, we are tasked with preparing the next generation for challenges and opportunities that lie ahead. One such challenge—and also an opportunity—is sustainable agriculture. Within this broad subject, crop diversification stands out as an essential strategy, with far-reaching implications for ecology, economy, and social equity. In this chapter, we will delve into the multi-faceted world of crop diversification, providing you with the foundational knowledge, teaching resources, and real-world examples needed to bring this critical subject to your classrooms.

The aim is not just to discuss what crop diversification is, but to examine how it can be effectively implemented and why it matters in a modern sustainable agricultural context. By the end of this chapter, you will be equipped with a comprehensive understanding of crop diversification, along with practical tools and methods to inspire and educate your students. We will explore the subject from its scientific underpinnings to its socio-economic benefits, making sure you are well-armed to guide your students in understanding and appreciating the role of crop diversification in sustainable agriculture.

What is Crop Diversification?

Crop diversification refers to the practice of cultivating multiple types of crops within a specific agricultural area, as opposed to focusing on a single crop, known as a monoculture. This approach has its roots in traditional agricultural systems and has gained renewed interest in recent years as a way to make agriculture more sustainable and resilient. It can manifest in various forms—crop rotation, intercropping, and polycropping are just a few examples.

In crop rotation, farmers plant different types of crops in succession on the same piece of land, usually in a planned sequence. This has multiple benefits, such as breaking the life cycles of pests, diseases, and weeds that are specific to one crop. It also allows different crops to utilize and contribute to the soil's nutrient profile differently, thereby promoting soil fertility.

Intercropping involves growing two or more different crops simultaneously in proximity, typically to exploit synergies between them. For instance, growing legumes alongside cereal crops can improve nitrogen fixation, benefiting both crops. Polycropping is similar but generally involves more than two crops and is not necessarily as planned or systematic as intercropping.

This diversified approach is in stark contrast to monocultures, where the cultivation of a single crop dominates vast expanses of land. Monocultures may offer high yields and operational simplicity but come with a host of environmental risks, including increased vulnerability to pests and diseases, soil degradation, and an elevated dependency on chemical fertilizers and pesticides.

Crop diversification can offer a range of benefits that contribute to the sustainability and resilience of agricultural systems:

- Biodiversity: Introducing different crops increases the biological diversity of the farm ecosystem, attracting a wider range of beneficial insects and microbes.
- Soil Health: Different crops have different nutrient requirements and offer different benefits to the soil. For example, legumes can fix nitrogen, improving soil fertility for future crops.

- Pest Management: The diversity of crops makes it more challenging for pests to establish themselves, reducing the need for chemical pesticides.
- Climate Resilience: A diversified farm is generally more resilient to variable weather conditions. If one crop fails due to a climate anomaly, others may still thrive, providing some level of income and food security.
- Economic Benefits: Crop diversification can open up new markets and revenue streams for farmers, reducing the economic risks associated with dependency on a single crop.

By embracing crop diversification, farmers stand to improve not just the environmental sustainability of their operations, but also their economic resilience and viability. It allows agriculture to move from a linear, reductionist model to a more circular, holistic one, thereby aligning it more closely with the principles of sustainability.

Benefits of Crop Diversification

The advantages of crop diversification extend far beyond just breaking the monotony of a single-crop system. From enhancing biodiversity to risk mitigation and improving soil fertility, the benefits are multifaceted and can be critical for the sustainability and resilience of agricultural systems. Here are some of the primary benefits:

- Enhanced Biodiversity: Crop diversification contributes to a more varied and robust ecosystem. A mix of crops invites a more diverse set of fauna, from beneficial insects to pollinators, thus promoting ecological balance. This can be especially crucial in today's context where loss of biodiversity is a major concern.
- Improved Soil Health: Different crops have varying nutrient requirements and growth patterns, which can complement each

other when grown in succession or proximity. For example, legumes fix nitrogen, benefiting subsequent non-leguminous crops. This not only enhances soil fertility but also reduces the need for synthetic fertilizers, which have long-term detrimental effects on soil structure and water quality.

- Pest and Disease Management: In a diversified cropping system, the life cycles of pests and diseases that are specific to one crop are disrupted, naturally reducing their presence. As a result, the dependency on chemical pesticides and herbicides decreases, leading to both economic and environmental gains.

- Reduced Economic Risk: Dependency on a single crop can be financially perilous if that crop fails due to disease, pests, or adverse weather conditions. Diversification serves as a form of insurance, allowing farmers to spread risk across multiple crops. If one fails, the others may still provide some revenue and food security.

- Climate Resilience: With the increasing uncertainties brought about by climate change, crop diversification offers a safety net. Some crops are more tolerant to extreme conditions such as drought or flooding, providing a buffer against unpredictable weather patterns. This is particularly important for smallholder farmers who may not have the means to invest in high-tech solutions for climate adaptation.

- Resource Optimization: Growing different crops can result in more efficient use of farm resources. For example, a taller crop can provide shade for a lower-growing, shade-tolerant crop, optimizing the use of space. Similarly, different crops have different water and nutrient needs, allowing for more efficient irrigation and fertilization schedules.

- Enhanced Nutritional Output: A diversified farm can produce a range of food products with different nutritional profiles. This not only benefits consumers by offering a more balanced diet but can also be more fulfilling for farmers who take pride in contributing to community health.

- Market Opportunities: Diversification can also open up new avenues for market engagement. For instance, a farmer traditionally growing only cereals could attract new customers by introducing legumes or vegetables into their crop rotation, diversifying their product offerings.
- Promotion of Sustainable Practices: By its very nature, diversification often promotes more sustainable farming practices, such as reduced tillage, cover cropping, and IPM, further contributing to the long-term health of the land and its surrounding environment.

In conclusion, crop diversification offers an array of benefits that contribute to the economic, ecological, and social sustainability of agriculture. By understanding and implementing these diverse systems, farmers and policymakers alike can work towards a more resilient and sustainable agricultural future.

How to Implement Crop Diversification

Implementing crop diversification can seem like a daunting task, especially for those who have traditionally practiced monoculture. However, with careful planning and the right resources, the transition can be smooth and rewarding. Here are some steps and considerations that guide this transformation:

1. Assess Your Resources: Conduct a thorough evaluation of your available resources—land, water, machinery, and labor. Knowing your resource capacity can help you make informed decisions about which crops to diversify into.
2. Local Climate and Soil Analysis: Understand the climatic conditions and soil type of your region. Certain crops thrive

better in specific conditions, and choosing those will increase your chances of success.

3. Market Research: Assess the demand for different crops in your local market or for export. High-value crops that have a good market can be more profitable even if they require more intensive management.

4. Consult with Experts: Before making any drastic changes, it's a good idea to consult with agricultural experts, extension officers, or even experienced farmers. They can provide insights that can help you avoid common pitfalls and offer advice tailored to your specific situation.

5. Crop Selection: Based on your resource assessment, climate and soil analysis, and market research, decide on a mix of crops that will be best suited for your farm.

6. Rotation and Interplanting Plans: Develop a detailed plan for how the crops will be rotated through fields or how they will be interplanted in the same field. Make sure to consider factors like differing nutrient requirements, growth cycles, and pest resistance.

7. Financial Planning: Create a budget, taking into account the costs of new seeds, fertilizers, pesticides (if you are not going organic), and additional labor, if required.

8. Implement Sustainable Practices: When diversifying, it's an excellent opportunity to integrate sustainable practices such as composting, organic farming, or permaculture methods.

9. Pilot Testing: Before scaling up, consider starting on a smaller piece of land to test the viability of your crop diversification plan. This can help you adjust your approach based on actual performance rather than theoretical assumptions.

10. Monitoring and Adjustment: Regularly monitor crop health, yields, and financial returns. Use this data to make necessary adjustments to your diversification strategy.

11. Education and Training: Ensure that everyone involved in the farming operations understands the new practices and protocols. This includes not only field workers but also administrative staff who may need to manage new types of contracts, sales channels, or data collection.
12. Community and Peer Learning: Engage with other farmers who are also practicing diversification. Learning from each other's experiences can provide valuable insights.

Implementing crop diversification is not a one-time activity but a continuous process of learning and adaptation. However, the myriad benefits from economic stability to environmental sustainability make the effort well worth it. With careful planning and execution, crop diversification can significantly elevate the resilience and profitability of your farming operation.

Case Studies: Crop Diversification

Lakewood Farms, Wisconsin, USA

Lakewood Farms in Wisconsin switched from a monoculture corn system to a diverse crop rotation system including soybeans, alfalfa, and winter wheat. Not only did they observe a 20% increase in corn yields, but they also noted a significant reduction in soil erosion and water runoff. The diversified approach reduced the need for fertilizers by 15%, resulting in significant cost savings.

Mysuru Model, Karnataka, India

In Karnataka, a group of farmers collaborated to diversify their traditionally rice-focused fields with vegetables, legumes, and even fish ponds. They implemented multi-tier cropping, where shorter plants are grown alongside taller plants to maximize the use of space. This dramatically increased yields and also offered the community a buffer against failed crops. The success of this model has inspired the regional government to promote crop diversification as a strategy for sustainable farming.

Oromia Region, Ethiopia

Farmers in this drought-prone area decided to incorporate drought-resistant crops like millet and sorghum into their farming systems. By diversifying from the traditional maize cultivation, these farmers managed to harvest crops even in dry seasons. This not only improved food security but also provided an additional income stream, thereby reducing the economic vulnerabilities of these communities.

Quechua Farms, Peru

In Peru's Andean region, small farmers practice a form of crop diversification called "Chacra," a traditional Andean agroforestry system. This involves planting various root crops, grains, and vegetables along with fruit trees. The system effectively controls pests and diseases while improving soil fertility and water retention. A study showed that these diversified plots were 20% more profitable than single-crop plots.

Niigata Prefecture, Japan

In this coastal region, farmers have successfully implemented a rice-fish culture. They raise fish in the same fields where rice is grown. This practice has not only led to increased rice yield but also provides an additional source of income through fish sales. The fish also act as natural fertilizers and help control pests, reducing the need for chemical inputs.

These real-life cases exemplify the multifaceted benefits of crop diversification—from improving yields and soil health to economic benefits and climate resilience. Each case study demonstrates how farmers, both small and large, can significantly benefit by incorporating diversity into their agricultural practices. These successes can serve as a guiding light for farmers considering transitioning from monoculture to diversified cropping systems.

Summary

Chapter 4 delved into the comprehensive understanding of crop diversification as a core practice in sustainable agriculture. Starting from its basic definition, we examined how diversifying crops is more than just planting different types of plants; it's a strategic approach to optimize land use, improve soil health, and mitigate risks like pest infestations and climate change impacts.

We discussed the multi-dimensional benefits of crop diversification, which extend beyond agricultural yield improvements to include ecological and economic advantages. Techniques for implementing diversification, such as crop rotation, intercropping, and multi-tier cropping, were detailed, offering practical insights into adopting these practices effectively.

Real-world case studies from around the globe, such as Lakewood Farms in Wisconsin, Mysuru Model in India, and Quechua Farms in Peru, demonstrated the success of diversification efforts. These instances served as tangible evidence of how diversified farming can be a win-win for both the environment and the farmers.

As a take-home message, crop diversification stands as a versatile solution to many of the challenges faced by modern agriculture. Its adoption could be a pivotal step towards the broader goal of sustainability, offering a roadmap for farmers, policymakers, and stakeholders in the agricultural sector.

Chapter 5: Pest Management

Welcome to Chapter 5, dedicated to the essential topic of Pest Management within the context of sustainable agriculture. As educators, understanding the nuances of pest management is crucial for conveying its importance and practical applications to your students. In this chapter, we will guide you through the diverse aspects of sustainable methods to control pests, contrasting them with traditional chemical-based methods. Our objective is to equip you with the educational tools and resources you need to incorporate these critical concepts into your curriculum effectively.

You will find a range of material here, from the scientific principles underlying pest management to classroom-friendly lesson plans and exercises. We'll look at alternatives to chemical pesticides, such as biological control methods, and discuss how these topics can be made engaging for students of various age groups. By the end of this chapter, you'll be well-prepared to teach your students about the complexities of pest management in a way that not only informs but also inspires them to think critically about sustainable solutions.

What is Pest Management?

Pest management is an integral aspect of sustainable agriculture and a subject that often stirs interest among students for its blend of science, technology, and practical application. At its core, pest management is about the strategies and methods employed to control or eradicate pests that damage or interfere with crops, livestock, and agricultural productivity. Pests may include insects, weeds, rodents, fungi, and various other organisms that cause harm to plants and animals.

Historically, the solution to dealing with pests was relatively straightforward: apply chemical pesticides. However, this approach often led to unintended environmental consequences such as soil degradation, water contamination, and harm to non-target species including humans. Additionally, the excessive use of chemical pesticides can lead to resistance in pests, making them more difficult to control in the future.

In today's more nuanced view, pest management is often seen through the lens of IPM, a multifaceted approach that seeks to manage pests in the most economical means while having the least possible impact on people and the environment. IPM involves a variety of strategies:

1. Monitoring: The first step is to understand what kinds of pests are present, and in what numbers. Tools for this might include traps, visual inspections, and population counts.
2. Identification: Not all insects or organisms in an agricultural setting are harmful. Students should learn to identify beneficial organisms, as they play an important role in natural pest control.
3. Threshold Levels: IPM teaches that action should only be taken when pest populations reach a certain level where the economic damage they cause justifies the cost of control measures.
4. Prevention: This can involve crop rotation, improved sanitation, and habitat manipulation to make the environment less hospitable to pests.
5. Control Methods: When pest levels reach a critical threshold, various methods, from chemical controls to biological solutions, are considered. Students can study how predatory insects, pheromones, biopesticides, and other innovative approaches are used alongside traditional pesticides.
6. Evaluation: After control methods are applied, ongoing monitoring and assessment are essential. This teaches students the value of adaptability and continual learning in agricultural practices.

7. Environmental Impact: In every step of IPM, consideration is given to the potential environmental impact of the chosen methods. It's a topic that offers great discussion material on ethics, sustainability, and long-term planning.

The subject of pest management opens up a plethora of opportunities for classroom activities, from lab experiments on the effectiveness of natural repellents to data analysis on pest population dynamics. In a more advanced setting, students can even design their own IPM programs based on hypothetical or real-life case studies, making the learning experience highly interactive and applied.

Understanding the intricacies of pest management not only enriches the academic experience but also equips students with essential skills for a career in modern, sustainable agriculture. Whether it's through practical application or through evaluating the ethical implications, teaching pest management provides a comprehensive view of a topic that is vital for the future of agriculture.

Biological vs. Chemical Methods

Biological and chemical methods represent two contrasting yet interconnected approaches in pest management that can spark engaging discussions and experiments in the classroom. Both have their advantages and disadvantages, and the choice between the two often depends on various factors such as the type of pest, scale of infestation, and environmental considerations.

Chemical Methods

Chemical methods involve the use of pesticides, herbicides, or fungicides to control or eradicate pests. They are typically quick-acting and can cover large areas, making them suitable for commercial agriculture. Here are some points that could facilitate classroom discussions and investigations:

- Effectiveness: Chemical methods are generally highly effective in reducing pest populations in the short term.
- Cost: The upfront cost of chemical pesticides can be lower than implementing biological controls, making it an economically attractive option for some farmers.
- Ease of Use: Chemicals can be easily applied through various methods, including spraying and dusting, requiring less specialized knowledge.
- Environmental Impact: The downside is that chemical methods can be harmful to the environment, affecting soil quality, water resources, and non-target species.
- Resistance: Overuse of chemicals can lead to pests developing resistance, requiring stronger or different chemicals for future control, a phenomenon that is a crucial teaching point in the dynamics of pest management.

Biological Methods

Biological methods use natural predators, parasites, or pathogens to control pests. This approach is more aligned with sustainable agricultural practices and can be a compelling area for student projects:

- Targeted Approach: Biological controls are often species-specific and less likely to harm non-target organisms. Students can explore how this specificity works through studies on predator-prey relationships.
- Sustainability: They are renewable resources that can maintain their population and continue to control pests over time without additional input.
- Cost-Efficiency: While the initial setup can be costly, the long-term maintenance costs are generally lower.

- Balance: Biological methods help maintain the ecological balance and can work in harmony with the natural environment.
- Limitations: These methods can be slow to act and often require a deeper understanding of the ecosystem for effective implementation.

Classroom activities could involve comparing the impact of chemical and biological methods on soil health, or analyzing data on the long-term costs and benefits of each approach. For a more hands-on experience, students could engage in experiments that use simple biological methods like introducing ladybugs to control aphids in a controlled environment.

In summary, chemical and biological methods each have their unique sets of pros and cons. Teaching these approaches side-by-side allows students to better grasp the complexities of modern pest management, enabling them to make more informed decisions in their future careers or academic pursuits. This nuanced understanding is invaluable in a world that increasingly values sustainable agricultural practices while grappling with the challenges of large-scale food production.

Natural Pesticides and Their Applications

Natural pesticides offer an alternative to synthetic chemical pesticides, providing an engaging area for educational exploration and classroom activities. These pesticides are derived from plants, minerals, or other organic sources and have been used for centuries in traditional farming practices. Let's dive into various types of natural pesticides and how they can be applied in sustainable agriculture.

Types of Natural Pesticides

- Botanicals: Extracts from plants like neem, pyrethrum, and garlic serve as potent natural pesticides. Neem oil, for example, disrupts the hormonal systems of insects and mitigates infestations.

- Mineral-Based: These include substances like diatomaceous earth and lime sulfur. Diatomaceous earth acts as a mechanical pesticide, dehydrating insects by absorbing their protective waxy coatings.
- Microbial: Bacteria, fungi, and viruses can serve as biological pesticides. Bacillus thuringiensis (Bt) is a commonly used bacterial pesticide that targets caterpillars and other leaf-eating insects.
- Homemade Mixtures: Simple recipes combining ingredients like vinegar, soap, and essential oils can create effective repellents for smaller-scale applications.

Applications

- Spraying: The most common method, suitable for both small gardens and larger fields. For classroom purposes, students can prepare their own natural pesticide sprays and apply them to test plants, noting the effects.
- Soil Treatment: Some natural pesticides, such as nematodes, are applied directly to the soil to control ground-dwelling pests. This offers an excellent opportunity for hands-on soil experiments.
- Fumigation: Certain plant extracts like eucalyptus can be used for fumigating an area, providing a broader coverage but often requiring specialized equipment.
- Insect Traps: Using substances like apple cider vinegar or pheromones to attract and trap pests can be a low-tech but effective method.

Advantages and Disadvantages

- Safety: Natural pesticides are generally safer for humans and pets, making them an excellent topic for classroom discussions about health and environmental safety.
- Environment: These pesticides are often biodegradable and less harmful to ecosystems. An exploration into the environmental impact of natural versus synthetic pesticides can offer insightful lessons.
- Resistance: Natural pesticides are generally less prone to inducing resistance in pests. Classroom activities could examine why this is the case through the lens of evolutionary biology.
- Cost and Availability: Although some natural pesticides may be more expensive, many can be made at home from common ingredients. A practical class could involve making homemade pesticides from everyday items.
- Efficacy: Natural pesticides often act more slowly and may require frequent application. Students can be encouraged to run experiments to test the efficacy of natural pesticides over time, comparing the results with synthetic counterparts

Natural pesticides, with their varied types and applications, offer a multidimensional approach to pest management that aligns well with educational objectives. They serve as an excellent tool for teachers aiming to instill a nuanced understanding of sustainable agriculture and eco-friendly practices in their students.

Case Studies: Pest Management

These real-world case studies provide a valuable resource for educators aiming to incorporate practical examples into their teaching on sustainable pest management. Each case highlights a unique approach, its challenges, and its outcomes, offering both a narrative and quantitative understanding of the subject:

- IPM in Apple Orchards - Washington State:
  - Scenario: A group of apple farmers in Washington faced severe pest issues, affecting both the quality and quantity of their harvest.
  - Method: IPM was implemented, combining chemical, biological, and mechanical methods.
  - Outcome: Within two seasons, there was a 40% reduction in chemical pesticide use and a 20% increase in yield.
  - Educational Value: This case can be used to discuss the IPM model and how diversified strategies can improve both economic and environmental outcomes.
- Use of Neem Oil in Organic Farms – India:
  - Scenario: Organic farmers in southern India were battling a surge of aphids and whiteflies.
  - Method: Neem oil, a natural pesticide, was sprayed on the crops.
  - Outcome: The infestation was controlled with minimal environmental impact and no harm to beneficial insects.
  - Educational Value: A discussion point for the role of botanicals in pest management and how ancient methods can be scientifically validated.
- Predatory Insects in Greenhouses – Netherlands:
  - Scenario: Tomato growers in the Netherlands were plagued by red spider mites.
  - Method: Introduction of predatory mites to the greenhouses.
  - Outcome: The predatory mites successfully managed the spider mite population, eliminating the need for chemical sprays.
  - Educational Value: This offers a hands-on classroom activity idea where students can set up mini-ecosystems to study predator-prey relationships.

- Community Engagement in Pest Management – Brazil:
  - o Scenario: Smallholder farmers in Brazil were grappling with caterpillar infestations in their cornfields.
  - o Method: Farmers were educated on the use of Bacillus thuringiensis (Bt), a bacterial pesticide, through community workshops.
  - o Outcome: The community saw a significant decrease in pesticide costs and an increase in yield.
  - o Educational Value: The case can serve as a talking point on the importance of education and community engagement in sustainable agriculture.
- Diatomaceous Earth in Grain Storage – Kenya:
  - o Scenario: Grain stores in rural Kenya were being destroyed by weevils.
  - o Method: Grain was stored with diatomaceous earth, a natural, mineral-based pesticide.
  - o Outcome: Weevil infestation was reduced by 90%.
  - o Educational Value: This case can introduce the concept of post-harvest pest management and how a simple mineral can have significant impact.

These case studies enrich the classroom experience, providing concrete examples of various pest management strategies in action. They offer the opportunity for critical thinking, problem-solving exercises, and can serve as the basis for student projects exploring the effectiveness and sustainability of different pest management options.

Summary

Pest Management is an integral aspect of sustainable agriculture, aiming to control pests in a manner that is efficient, economical, and environmentally sound. This chapter explored the various methodologies in pest management, spanning from traditional chemical methods to more sustainable approaches like biological control. It shed light on the fundamental concept of pest management, emphasizing the need for a multi-pronged strategy.

We dove deep into comparing biological versus chemical methods, outlining the pros and cons of each. The focus was on striking a balance between immediate effectiveness and long-term sustainability. We also explored natural pesticides like neem oil and Bacillus thuringiensis (Bt), demonstrating how these can offer safer alternatives to chemical pesticides.

The chapter was enriched with real-world case studies, ranging from IPM in Washington's apple orchards to the use of predatory insects in Dutch greenhouses. These cases provided practical insights into how sustainable practices are being implemented globally, offering inspiration and models for adaptation.

The aim of this chapter has been to equip educators with a comprehensive understanding of pest management, backed by science and practice. The case studies serve as valuable tools to make the subject relatable and engaging for students, facilitating more effective teaching and learning.

Chapter 6: Livestock Management

In the vast tapestry of sustainable agriculture, livestock management often holds a central place, both as a challenge and an opportunity for farming communities. Traditionally, livestock has been a crucial part of agriculture, providing not just meat and dairy but also aiding in nutrient cycling, land management, and even as draft power. However, contemporary practices in livestock management have sometimes overlooked sustainability, leading to environmental degradation, resource depletion, and ethical concerns. This chapter aims to bridge that gap by providing educators with a comprehensive framework to teach the intricacies of sustainable livestock management.

We will explore key concepts, from understanding the multifaceted roles that livestock play in agroecosystems to the latest trends in sustainable practices. Whether it's rotational grazing, integrated farming systems, or alternative feed sources, this chapter provides a panoramic view of viable options. This resource-rich guide, embellished with real-world case studies, will not only inform but also equip teachers with the tools to bring about meaningful discussion and change in their classrooms. Together, we'll delve into the future of livestock management, where sustainability is not just a buzzword but a guiding principle.

Importance of Livestock in Agriculture

Understanding the importance of livestock in agriculture is crucial for grasping the full scope of sustainable farming practices. For centuries, livestock have been at the core of agricultural systems, offering multiple benefits that go beyond the primary products of meat, milk, and eggs. Livestock are essential players in nutrient cycling, converting plant materials that are indigestible for humans into valuable food products, while their waste contributes organic matter to improve soil fertility.

In many parts of the world, livestock serve as a form of natural capital, providing economic security and livelihoods for farming communities. They are often considered 'walking bank accounts,' offering financial flexibility through their sale or trade. Livestock are not just income generators; they also contribute to risk diversification. When crop failure occurs due to adverse weather conditions or pest attacks, livestock serve as a buffer, safeguarding the economic stability of the household.

The symbiotic relationship between livestock and land use further showcases their importance. Through grazing, animals contribute to natural land management by controlling plant overgrowth and reducing fire risks. Rotational grazing, a practice where livestock are moved between different paddocks or fields, can even help in reclaiming degraded land by promoting the growth of native vegetation.

Livestock also contribute to sustainable agriculture through integrated farming systems. These are systems where the waste from one component serves as a resource for another. For instance, animal manure can be used to fertilize crops, while crop residues can serve as animal feed. Such a circular system reduces waste and optimizes resource use, making agriculture more efficient and less dependent on external inputs.

As we move toward a more sustainable future, the role of livestock in agriculture becomes even more critical. They can be integral to closed-loop systems that recycle nutrients and improve soil health. New practices in livestock management are continually being developed to reduce their environmental footprint, such as precision feeding to reduce methane emissions or alternative organic feeds to replace conventional grain. By educating about these aspects, teachers can help students understand the complex role of livestock in agriculture and the innovative solutions that are being developed to make livestock rearing more sustainable.

Thus, livestock are not merely subsidiary to agriculture; they are a cornerstone of a multifaceted system that contributes to environmental health, economic viability, and food security. Their significance underscores the need for sustainable management practices that benefit not just the livestock but the entire agricultural system.

Sustainable Livestock Practices

Sustainable livestock practices are gaining attention as a way to balance the nutritional, economic, and ecological aspects of agriculture. It is crucial to move away from conventional methods that are often resource-intensive and environmentally damaging. Teachers can guide students in exploring these sustainable alternatives, which are not just trends but necessities for the future of farming.

One cornerstone of sustainable livestock practices is rotational grazing. This method involves dividing pastures into smaller sections and rotating livestock through them. This technique allows the grass to recover, promoting root growth and reducing soil erosion. It also helps control parasite cycles, reducing the need for chemical dewormers. Rotational grazing optimizes the use of pasture and reduces the need for supplemental feed, which can be both expensive and carbon-intensive.

Another critical aspect is the use of locally adapted breeds. These are livestock breeds that have evolved or have been bred to be well-suited to local environmental conditions. They are often more resistant to local diseases and better adapted to local feeds and climate, reducing the need for external inputs like medications or specialized feeds.

Animal welfare is an essential part of sustainability. Happy, healthy animals are more productive and less resource-intensive. Sustainable livestock practices thus also involve creating suitable living conditions, which means adequate space, natural light, and opportunities for natural behaviors. Implementing animal welfare standards not only aligns with ethical considerations but also often results in improved meat, milk, and egg quality.

Nutrient recycling is another key principle. In integrated crop-livestock systems, the manure from livestock can be used to fertilize fields, reducing the need for synthetic fertilizers. This approach turns a waste product into a valuable resource, contributing to soil health and reducing water pollution. Manure management, when done correctly, can significantly reduce the emission of greenhouse gases like methane and nitrous oxide.

Lastly, precision livestock farming uses technology to monitor the health and well-being of individual animals. For instance, wearables can provide real-time data on an animal's physiological state, allowing for more timely and targeted care. This not only enhances animal welfare but can also lead to resource savings by optimizing feeding and healthcare interventions.

By instilling these principles and practices in the next generation, educators can play a critical role in the shift towards more sustainable livestock management. The shift is not just about making livestock farming more eco-friendly; it's about creating a system that is resilient, ethical, and sustainable for the long run. Teaching these practices offers students a holistic understanding of agriculture, making them better equipped to contribute to a more sustainable future.

The Role of Livestock in Agroecosystems

Livestock play a multifaceted role in agroecosystems, contributing not only to food production but also to nutrient cycling, land management, and biodiversity. Educators aiming to impart comprehensive knowledge about sustainable agriculture should not overlook the importance of livestock in agroecosystems. By exploring the interrelationships between livestock and their environment, students gain a more nuanced understanding of how sustainable farming practices can be integrated for holistic land management.

One of the most critical roles of livestock in agroecosystems is nutrient cycling. The manure produced by animals is rich in essential nutrients like nitrogen, phosphorus, and potassium. When properly managed, this manure can be used as organic fertilizer for crops, reducing the need for synthetic inputs and enhancing soil fertility. This creates a synergistic relationship between crop and livestock farming, as crop residues can also serve as feed for animals.

Livestock also play a crucial role in land management. Through practices like rotational grazing, animals can help control weed populations and reduce the risk of fire by consuming dry vegetation. Furthermore, the hoof action of grazing livestock can break up compacted soils, allowing for better water infiltration. This not only makes the land more resilient to drought but also prevents water runoff and soil erosion. Educators can incorporate this aspect into the curriculum by designing projects or field visits that show these principles in action.

Moreover, livestock can contribute to biodiversity in agroecosystems. For instance, certain livestock breeds are well-adapted to local conditions and can thrive without high levels of external inputs. Preserving these breeds can conserve genetic diversity, which is crucial for resilience against diseases and changing climate conditions. Students could benefit from lessons on the importance of genetic diversity, perhaps even visiting local farms that practice sustainable livestock management with native breeds.

In addition, livestock can be integrated into agroforestry systems, where they can graze on fallen leaves or fruits, converting these into valuable manure. This form of silvopastoral system represents a win-win situation, enhancing both tree and animal productivity while improving ecological resilience. The topic offers an excellent opportunity for educators to discuss the concept of ecosystem services, from carbon sequestration to habitat creation for wildlife.

Finally, livestock contribute to the economic viability of farming systems. Products like milk, meat, and wool provide farmers with diverse income streams, making them less vulnerable to market fluctuations in crop prices. For educational purposes, this aspect serves to illustrate the interconnectedness of ecological and economic sustainability in agricultural systems.

To sum up, the role of livestock in agroecosystems extends far beyond mere food production. They are pivotal in nutrient cycling, land management, biodiversity conservation, and economic sustainability. By teaching students about these interrelationships, educators equip them with a broader, more holistic understanding of sustainable agriculture.

Case Studies: Livestock Management

For educators aiming to instill a nuanced understanding of sustainable agriculture, real-world case studies can be invaluable. Below are two case studies that showcase sustainable livestock management practices, providing concrete examples for the theoretical principles discussed in the classroom.

Case Study 1: Rotational Grazing in New Zealand

One notable example is a sheep and cattle farm in New Zealand that implemented a rotational grazing system. Instead of allowing livestock to graze freely over large areas, the farm divided its pastures into smaller paddocks. Animals were moved between these paddocks regularly, allowing vegetation in grazed paddocks to recover:

- Results: This strategy led to healthier soils, fewer weeds, and increased productivity. Soil tests showed higher levels of organic matter and nutrients, and the need for synthetic fertilizers decreased dramatically. Furthermore, the farm reported higher profits, as healthier animals led to higher yields of meat and wool.
- Educational Takeaway: The case study could serve as a basis for class discussions on soil health, land management, and economic sustainability. Students could analyze the data and evaluate the benefits and drawbacks of rotational grazing.

Case Study 2: Integrated Crop-Livestock Systems in Brazil

In Brazil, a farm integrated crop and livestock farming by planting soybean and maize alongside pasture areas for cattle. During the off-season for crops, cattle were allowed to graze on the stubble and crop residues, thereby enhancing soil fertility through their manure:

- Results: Crop yields on this farm increased by 20%, while methane emissions from cattle decreased due to a more varied and nutritious diet. Moreover, the farm needed fewer chemical inputs, reducing both costs and environmental impacts.
- Educational Takeaway: This case study presents an excellent example of how integrating crops and livestock can lead to win-win situations for both environmental sustainability and farm profitability. Students can engage in activities like cost-benefit analyses or debates on the scalability of such integrated systems.

Both case studies serve as practical illustrations of how sustainable livestock management can result in multiple benefits, from soil health to economic gains. These real-world examples can enrich the educational experience, enabling students to better understand and appreciate the complexities and potential advantages of incorporating livestock into sustainable agricultural systems.

Chapter 7: Technology in Sustainable Agriculture

In an era where technology is rapidly advancing, its application in agriculture is not only inevitable but also immensely beneficial. Chapter 7 focuses on how technology plays a pivotal role in sustainable agriculture, offering solutions that can make farming practices more efficient, less resource-intensive, and more environmentally friendly. From drone technology for crop surveillance to sophisticated irrigation systems that optimize water use, technology has provided a myriad of tools for the modern farmer.

However, the incorporation of technology in agriculture is not without challenges and ethical considerations. Issues like accessibility, affordability, and the impact on employment are crucial when considering technological solutions for sustainable farming. Furthermore, there's the question of whether technology can sometimes do more harm than good, especially when not managed sustainably.

This chapter aims to provide educators with an in-depth understanding of the different types of technologies available for sustainable agriculture. It will also delve into the advantages, limitations, and ethical considerations that come with integrating technology into farming practices. The ultimate goal is to equip teachers with the knowledge and resources they need to engage students in meaningful discussions and activities centered around the use of technology in achieving a more sustainable future for agriculture.

Types of Technology in Agriculture

When discussing the role of technology in sustainable agriculture, it's essential to understand the varied types of technologies that have come to define modern farming practices. These technologies span from simple hand-held devices to complex machinery and software systems:

- Precision Agriculture: One of the most significant advancements in farming is precision agriculture, which utilizes GPS technology, sensors, and data analytics to optimize field-level management with regard to crop farming. It allows farmers to apply inputs like fertilizer and irrigation water more efficiently, thereby saving resources and increasing yields.
- Drones: Unmanned Aerial Vehicles (UAVs), commonly known as drones, are used for various agricultural tasks like planting, crop spraying, and aerial mapping. They offer a quick and relatively inexpensive way to survey large plots of land and can be integrated with software that can analyze crop health.
- Automated Tractors and Harvesters: Self-driving tractors equipped with advanced machine learning algorithms can perform tasks like planting, plowing, and harvesting more efficiently. They can operate round the clock, thereby increasing productivity.
- Internet of Things (IoT): Sensors placed in the field can send real-time data to farmers' smartphones or computers. These sensors can measure soil moisture, detect pests, and even monitor the health of livestock.
- Controlled Environment Agriculture (CEA): Technologies like hydroponics and aeroponics allow for the controlled growth of crops in indoor settings. These systems are particularly useful for growing crops in areas with adverse climatic conditions.
- Bioengineering: Although a topic of debate, GMOs have the potential to resist pests, tolerate harsh conditions, and provide higher nutritional content. However, the use of GMOs in agriculture is highly regulated and subject to ongoing ethical discussions.

- Mobile Apps: Numerous applications can help farmers monitor weather conditions, track commodity prices, and even diagnose plant diseases through image recognition.
- Vertical Farming: This involves growing crops in stacked layers, usually indoors, where all environmental conditions are controlled. Although energy-intensive, vertical farming has the advantage of producing food close to urban areas, reducing the carbon footprint associated with transportation.
- Blockchain: Although still in its nascent stage in agriculture, blockchain technology promises transparent and unchangeable ledgers, which could be particularly useful in tracing the origin of food products, thereby ensuring quality and safety.

- Livestock Technology: Innovations like RFID tags for animals make it easier to manage and monitor livestock health, while biometric systems can recognize individual animals based on unique features like retinal scans.

Understanding these diverse technologies is crucial for educators aiming to teach sustainable agriculture. Each technology offers its unique set of benefits and challenges, making it important for students to grasp the complexities involved in their application.

Benefits and Challenges of Technology

The integration of technology into agriculture is a double-edged sword, bringing about a host of benefits while also posing several challenges that need to be judiciously managed.

Benefits:

- Efficiency and Productivity: The foremost advantage of using technology is the increase in efficiency and productivity. Automation and precision farming techniques allow for more efficient use of resources like water, soil, and fertilizers. Drones and automated harvesters can cover larger areas in a shorter time, which is crucial for the timely planting and harvesting of crops.

- Data-Driven Decisions: With the advent of the Internet of Things (IoT) and machine learning, farmers can make data-driven decisions. Whether it's about the best time to sow seeds or irrigate fields, data analytics provides invaluable insights.

- Resource Conservation: Sustainable practices are more easily implemented with the right technology. Precision agriculture allows farmers to apply just the right amount of water, nutrients, and pesticides, which not only conserves resources but also reduces the environmental impact.

- Reduced Labor Costs: Automation drastically cuts down the need for manual labor, thereby reducing overall costs. This is particularly beneficial for farmers in developed countries where labor costs are high.

- Quality and Safety: Technologies such as blockchain can improve food safety by making the supply chain transparent. Consumers can trace the origin of the products they consume, reassuring them of the quality and safety of the food.

Challenges:

- Cost: One of the most significant barriers to the widespread adoption of agricultural technology is the initial investment required for machinery, software, and training. Not all farmers can afford to transition to tech-savvy farming.

- Skill Gap: The use of advanced technologies requires a level of skill and understanding that many traditional farmers may lack. Educational and training programs are essential but not always easily accessible.

- Environmental Impact: Some technologies, particularly those reliant on heavy machinery and chemical inputs, could have a detrimental impact on the environment if not managed sustainably. For example, automated tractors can compact soil, reducing its fertility over time.

- Data Security: As agriculture becomes more interconnected, it becomes more vulnerable to data breaches and cyber-attacks. Farmers must be educated about the importance of cybersecurity to protect their data and systems.
- Social and Ethical Concerns: Issues like job displacement due to automation and the ethical implications of genetically modified crops can't be ignored. Public opinion varies on these topics, which makes them difficult to navigate.

Understanding the benefits and challenges of integrating technology into agriculture is key for a balanced approach. Educators must equip students with the knowledge to maximize the advantages while mitigating the disadvantages. This nuanced understanding will be critical for the future leaders in the field of sustainable agriculture.

Case Studies: Technology in Action

The use of technology in sustainable agriculture is taking place all around the world, radically transforming traditional practices and resulting in more efficient and environmentally friendly operations. Here are a couple of real-world case studies that demonstrate the impact of technology in agriculture.

Case Study 1: Precision Agriculture in Iowa, USA

In the farmlands of Iowa, USA, a group of farmers teamed up with agricultural technology companies to implement precision agriculture. Using drones equipped with multispectral cameras, they were able to map their fields down to the square inch. This detailed mapping was then integrated into their irrigation and fertilization systems:

- Outcome: The result was a 20% reduction in water usage and a 15% reduction in fertilizer use. Crop yields also improved due to targeted irrigation and fertilization.
- Lesson: The integration of drone technology with existing agricultural practices can result in significant resource savings and increased yield.
- Scalability: The model has now been replicated across multiple states, validating its effectiveness and scalability.

Case Study 2: Mobile Apps for Small-Scale Farmers in Kenya

Small-scale farmers in Kenya often struggle with access to reliable information on weather, pest control, and best practices. To address this issue, a mobile application was developed that provides real-time, localized information to farmers via text messages:

- Outcome: The farmers who used the app reported a 30% increase in crop yields and were better able to manage pests and diseases, all without using harmful chemicals.
- Lesson: Simple technology solutions can empower small-scale farmers, improving their livelihoods and the sustainability of their farming practices.
- Accessibility: The low-cost nature of the app makes it accessible to even the most remote farmers, provided they have a basic mobile phone.

These case studies show that technology can indeed make a substantial difference in the field of agriculture, whether it's through high-tech solutions like drones and precision farming in developed countries, or through accessible, information-based applications in developing nations. However, it's crucial that the implementation of these technologies be carried out thoughtfully and sustainably, taking into account the needs of the community and the environment.

Chapter 8: The Economics of Sustainable Agriculture

Sustainable agriculture is not just an environmental or ethical consideration; it's a viable economic model that can redefine the future of farming and food production. The economics of sustainable agriculture involves more than just immediate gains; it also includes long-term profitability, equitable distribution of resources, and the creation of a resilient system that can withstand fluctuations in climate and market conditions. This chapter aims to provide an in-depth exploration of the economic factors that play a pivotal role in making agriculture more sustainable. We will look at the costs, benefits, challenges, and opportunities associated with sustainable farming practices. Topics will include the economic incentives for farmers to adopt sustainable methods, how sustainability can affect overall market dynamics, and the real-world applications where sustainable agriculture has proved to be economically viable. This economic perspective is crucial for teachers, students, and anyone interested in understanding the full spectrum of sustainable agriculture. By the end of this chapter, you will have a comprehensive understanding of why sustainable agriculture is not just good for the planet, but also for the wallet.

The Economic Viability of Sustainable Practices

In an era where climate change, soil degradation, and water shortages are everyday challenges, the agricultural industry faces increasing pressure to adopt sustainable practices. However, the transition often poses an essential question: Is sustainable agriculture economically viable? This is a complex issue, involving both short-term expenses and long-term gains. While initial investments in sustainable technologies and practices can be high, the economic benefits often manifest over time, offering cost savings, greater yield stability, and new market opportunities. This section delves deep into the intricacies of the economic factors shaping sustainable agriculture, examining the real costs and tangible benefits involved.

How Sustainability Affects Cost-Effectiveness

Transitioning to sustainable agriculture often comes with upfront costs, such as the adoption of new technologies, the purchase of organic inputs, or the undertaking of specialized training programs. While these initial expenditures can be a barrier for some, it's crucial to view them as investments in the farm's long-term viability. Implementing sustainable practices can significantly reduce recurring expenses. For example, a decreased reliance on synthetic fertilizers and pesticides cuts down operational costs while preserving soil health for future generations. Sustainable strategies like crop rotation and agroforestry offer additional economic advantages. These practices improve soil fertility and help in pest management, leading to better yields without the continual purchase of expensive chemicals. They also make the land more resilient, reducing the risks associated with environmental stresses and climate fluctuations. Thus, the cost-effectiveness of sustainable farming becomes apparent when evaluated over an extended period, taking into account both the reduction in input costs and the increased productivity of healthier lands.

Market Trends Supporting Economic Viability

Consumer demand is undeniably skewing towards sustainably-produced goods, offering a potent financial incentive for farmers to adopt sustainable practices. A growing segment of the population is willing to pay a premium for food that is organic, non-GMO, or sustainably sourced, thereby making these farming methods more economically viable. This trend is not just a fad; it is supported by increased awareness of the environmental and health impacts of conventional agriculture. As consumers become more educated, the market for sustainable products expands, attracting more retailers and creating a positive feedback loop that further boosts demand. This increasing market opportunity allows farmers to diversify their product offerings and cater to these niche but growing segments. Moreover, several certification bodies and labels now exist to verify sustainable practices, providing an additional marketing advantage for producers. Thus, the evolving consumer landscape is actively supporting the economic viability of sustainable farming, making it an opportune time for farmers to transition and tap into these profitable markets.

Profitability Analysis: Case Studies

Several real-life case studies substantiate the economic viability of sustainable farming practices. One notable example is the Rodale Institute's Farming Systems Trial, the longest-running comparison of organic and conventional agriculture in the U.S. The study showed that organic yields matched conventional yields but did so with 30% less energy, less water, and no pesticides. Over time, the organic systems have proven to be more profitable due to higher market prices and lower input costs.

Another case study comes from the experience of Jean-Martin Fortier, a farmer in Quebec, Canada. Fortier manages a 1.5-acre micro-farm that focuses on diversified vegetable growing. By utilizing intensive, sustainable farming methods, he has been able to generate over $140,000 in sales per acre annually. His book "The Market Gardener" describes his methods and provides a viable model for small-scale, sustainable agriculture.

In a different context, the Indian state of Sikkim went 100% organic in 2016, transforming its agriculture sector. The initial investment in training and organic certification was high, but farmers have reported better yields and higher profits in the long run, especially as their produce fetches higher market prices. The state has also seen a significant reduction in soil erosion and water pollution.

Lastly, a case in point from Africa is the push for agroforestry in Zambia. Farmers who integrated trees into their farms not only saw improvements in soil fertility but also gained additional income from tree products like fruits, nuts, and timber, thereby diversifying their income streams and enhancing their economic stability.

These real-life cases underscore the economic benefits of adopting sustainable practices, including higher yields, better soil health, and ultimately, greater profitability.

Government Subsidies and Grants

Financial support from governments can significantly tip the scale in favor of sustainable agriculture. Subsidies, grants, and low-interest loans are increasingly becoming available to farmers who are willing to transition to more sustainable practices. These incentives serve dual purposes: they help farmers manage the initial costs of shifting away from conventional methods, and they accelerate national goals towards more sustainable food production systems. This financial backing thus becomes a critical factor in enhancing the economic viability of sustainable agriculture, making it a more attractive and feasible option for farmers.

Challenges and Risks

While sustainable agriculture presents an optimistic picture, there are associated challenges and risks that cannot be overlooked. Transitioning from traditional to sustainable farming can result in initial lower yields, posing market risks. Additionally, sustainable ecosystems can be more sensitive to climatic fluctuations, exposing farmers to environmental risks. These challenges make risk assessment and mitigation an integral part of sustainable farming. Appropriate planning, diversification, and adaptation strategies are crucial to ensure that the shift to sustainable agriculture remains economically viable in the long term.

Conclusion and Future Outlook

Navigating the economics of sustainable agriculture is a multifaceted task, but the future looks promising. Driven by evolving consumer preferences and aided by government support, sustainable farming is increasingly proving its financial viability. This shift represents not just a trend, but a meaningful move toward a more sustainable and profitable future in agriculture.

Government Policies and Incentives

Government policy is a critical lever in shaping the landscape of sustainable agriculture. By offering financial incentives, establishing regulations, and funding research, governments can either accelerate or impede the transition to more sustainable farming practices. This section explores these aspects in depth.

Subsidies

Subsidies have long been a part of agricultural policy, traditionally focused on supporting large-scale, conventional farming methods. These subsidies often cover the costs of synthetic fertilizers, pesticides, and irrigation systems, thus incentivizing methods that may be unsustainable in the long run. Shifting the focus of these subsidies towards more sustainable farming practices can significantly impact the sector's transformation. Financial support for organic fertilizers, drought-resistant seeds, and water-efficient irrigation could drive a more widespread adoption of sustainable methods.

Start-up grants are another crucial aspect of government subsidies. Transitioning from conventional to sustainable farming requires initial investments in technology, soil restoration, and training. Offering small grants to farmers willing to initiate sustainable practices can substantially lower the financial barriers of such a transition. For instance, grants could cover the initial costs of soil testing, organic seed purchase, or the installation of solar-powered irrigation systems. By targeting these initial setup costs, governments can make sustainable farming more accessible and attractive for farmers who might otherwise be deterred by the upfront investment.

Tax Incentives

Tax incentives can serve as powerful tools for encouraging the adoption of sustainable agricultural practices. Green Tax Credits are one such mechanism, offering financial benefits to farmers who invest in environmentally friendly technologies or practices. For instance, a farmer who installs a rainwater harvesting system or uses energy-efficient machinery could be eligible for tax credits, reducing their overall tax liability. This not only makes these eco-friendly options more affordable but also provides a direct financial reward for sustainable behavior.

Depreciation benefits for sustainable infrastructure offer another avenue for financial incentives. Conventional tax systems allow businesses to write off the depreciation of assets over time. If governments were to accelerate these depreciation benefits for sustainable infrastructure, it would become more economically viable for farmers to invest in them. For example, a farmer investing in solar pumps or wind turbines for energy could be allowed to write off these investments at an accelerated rate, making these technologies financially more attractive. This could lead to increased investment in long-lasting, sustainable farm infrastructure, thereby making farms more efficient and eco-friendly in the long run.

Regulatory Measures

Regulatory measures have the potential to significantly impact the transition to sustainable agriculture. One such regulatory intervention could be the establishment and enforcement of stringent quality standards for sustainable products. For example, certification programs could be established to verify the sustainable practices behind a farm's produce, ensuring they meet set criteria such as organic, non-GMO, or fair trade. This not only helps consumers make informed decisions but also adds value to sustainably produced goods, making them more competitive in the market.

Zoning laws offer another avenue for governmental influence. The conversion of farmland to commercial or residential use is a growing concern that threatens local food production and sustainable agriculture. Governments can enforce zoning laws that protect farmland and designate certain areas solely for agricultural purposes. These laws can preserve the integrity of the land and ensure it is utilized for long-term food production rather than short-term financial gains from property development. By strategically implementing zoning laws, authorities can also control the encroachment of urban areas into prime agricultural lands, thus preserving them for future generations of farmers. Both quality standards and zoning laws can play pivotal roles in making sustainable agriculture more widespread and economically viable.

Research and Development

Research and Development (R&D) serves as the backbone for advancing sustainable agriculture. One critical component is the allocation of funding for sustainable farming techniques. Governments can support universities, agricultural institutes, and even independent research organizations to develop technologies and practices that are both sustainable and economically viable. Funding can help to accelerate innovation in areas like soil health, water conservation, and IPM, which are crucial for the long-term sustainability of agriculture.

Extension services represent another important aspect of R&D that can be backed by government intervention. Government-funded extension services can play an invaluable role in bridging the gap between research and its practical application in the field. These services, often offered through universities or specialized government agencies, can educate farmers on the benefits of sustainable practices, how to implement them, and how to adapt to evolving challenges like climate change. Extension services can offer hands-on workshops, field demonstrations, and ongoing support, empowering farmers with the knowledge and skills they need to transition successfully to sustainable methods. By focusing on these two key areas, Funding and Extension Services, governments can significantly contribute to making sustainable agriculture both a practical and economically viable option for farmers.

Case Studies: Economic Success Stories

Case Study 1: Organic Valley Cooperative, USA

Organic Valley, based in the United States, started in 1988 with a group of Wisconsin farmers dedicated to organic farming. Faced with falling milk prices, these farmers decided to take a sustainable approach. Fast forward to today, Organic Valley is a cooperative of over 1,800 farmers, making it one of the largest organic farming cooperatives globally. In 2019, the co-op reported over $1.1 billion in sales, proving that an organic approach is not just environmentally sound but also economically robust. Key to its success are government grants for organic certification and strong consumer demand for organic products.

Case Study 2: Alentejo Agroecological Transition, Portugal

In Portugal's Alentejo region, local farmers faced severe water scarcity and soil degradation. The European Union's Common Agricultural Policy (CAP) provided subsidies for transitioning to agroforestry, which combines agriculture with tree planting. This transition resulted in higher water retention and improved soil health. Economically, farmers benefited from diversified income sources, including cork production from the newly planted trees and increased resilience to market fluctuations.

Case Study 3: Zero-Budget Natural Farming, India

In the Indian state of Andhra Pradesh, the Zero Budget Natural Farming (ZBNF) program has transformed the agricultural landscape. Funded by both state and international agencies, ZBNF promotes low-input, chemical-free farming. As of 2020, the program has reached over half a million farmers. Those who adopted these practices have reported a 20-40% increase in income due to lower input costs and a rise in crop yields, in addition to environmental benefits such as improved soil health.

Case Study 4: Integrated Rice-Duck Farming, China

In southern China, traditional rice farming methods were resulting in high water usage and increased pest issues. Local farmers started using Integrated Rice-Duck Farming, a method that leverages ducks to eat pests and weeds. This significantly reduced the need for synthetic pesticides and fertilizers. Local government support and incentives facilitated this transition. Farmers who adopted this method saw an average increase of 20% in their yields and a significant reduction in input costs.

Case Study 5: Laikipia Permaculture Centre, Kenya

Laikipia Permaculture Centre in Kenya began as a community initiative to combat soil erosion and deforestation. With support from the Kenyan government and international organizations, the center transitioned to permaculture and sustainable livestock management. Farmers participating in the project reported a 35% increase in income due to higher yields and diversified crops like legumes and fruits, which also improved the nutritional intake of local communities.

These real-world case studies demonstrate the potential for economic success through sustainable agriculture, supported by well-placed governmental policies and market demand for sustainable products. They show that with appropriate strategies and support, sustainable agriculture can not only be environmentally beneficial but also economically viable.

Chapter 9: Teaching Sustainable Agriculture

Teaching Sustainable Agriculture is a critical component in the quest for a more eco-friendly, economically viable, and socially just food system. As the world faces escalating challenges like climate change, diminishing natural resources, and population growth, agriculture stands at the intersection of these global crises. The need to impart sustainable agricultural practices to the next generation of farmers, agricultural scientists, and policymakers has never been more urgent.

Education in this field goes beyond simply disseminating information; it aims to foster critical thinking, promote ethical considerations, and impart practical skills for sustainable farming. The purpose of this chapter is to explore the different avenues, methods, and curricula that can be used to effectively teach sustainable agriculture. We will delve into traditional educational settings, vocational training, community outreach programs, and online platforms that make sustainable agricultural education accessible to various audiences.

By fostering a deeper understanding and appreciation of sustainable farming methods, educators can cultivate a future generation that not only feeds the world but does so in a way that is harmonious with the planet. This chapter provides an in-depth look at how education in sustainable agriculture can be structured, the challenges it faces, and the impact it has on creating a more sustainable future.

Lesson Plans and Resources

One of the foundational elements for effectively teaching sustainable agriculture is the development of comprehensive lesson plans and resources. These instructional tools serve as the backbone for educational programs, outlining the knowledge and skills that students should acquire by the end of the course:

- Introduction to Sustainable Agriculture:
  - Objective: To provide students with a basic understanding of what sustainable agriculture is and why it is important.
  - Resources: PowerPoint presentations, videos, guest lectures from experts in the field.
  - Activities: Group discussions on sustainable vs. conventional agriculture, quizzes, and field trips to local sustainable farms.
- Soil Health and Management:
  - Objective: To educate students about the importance of soil health in sustainable farming.
  - Resources: Soil testing kits, literature on composting and soil nutrition, interactive online tools.
  - Activities: Hands-on soil testing, composting workshops, and in-class demonstrations on soil erosion.
- Water Conservation Techniques:
  - Objective: To teach students various water-saving methods applicable in agriculture.
  - Resources: DIY irrigation kits, videos on water conservation, case studies.
  - Activities: Building a miniature irrigation system, calculating water usage, and evaluating the effectiveness of different water-saving techniques.
- Crop Diversification and Rotation:
  - Objective: To understand the importance of growing a variety of crops and rotating them to improve soil health and reduce pest pressure.
  - Resources: Charts, graphs, research papers, and virtual simulations.

- Activities: Designing a crop rotation plan, analyzing data on the benefits of crop diversification, and role-playing activities.
- IPM:
  - Objective: To teach the principles of IPM and how it can be a more sustainable option than chemical pesticides.
  - Resources: Pest identification guides, chemical vs. natural pesticide comparison charts.
  - Activities: Identification of common pests and diseases, creating a natural pesticide, and evaluating its effectiveness.

- Sustainable Livestock Management:
  - Objective: To enlighten students about humane and sustainable methods of livestock farming.
  - Resources: Videos, interviews with sustainable livestock farmers, research papers.
  - Activities: Virtual farm tours, panel discussions, and debates on conventional vs. sustainable livestock management.
- Economics of Sustainable Agriculture:
  - Objective: To impart knowledge on the economic viability and benefits of sustainable agriculture.
  - Resources: Financial models, market research reports, and case studies.
  - Activities: Creating a mock business plan for a sustainable farm, analyzing the ROI of sustainable practices.

Educators must continually update these lesson plans and resources to include the latest research, technologies, and methodologies in the field. A wide variety of resources, ranging from textbooks to interactive apps and online platforms, should be utilized to cater to diverse learning styles. An effective lesson plan is flexible and adaptable, allowing educators to modify it according to the specific needs and interests of their students.

Incorporating Sustainability in School Curriculum

Incorporating sustainability into the school curriculum is not just a matter of adding a new subject or course; it's a paradigm shift that requires integrating the principles of sustainability across various academic disciplines. The goal is to instill a sense of responsibility and equip students with the practical skills and theoretical knowledge they need to make informed decisions that are socially, economically, and environmentally sustainable:

- Interdisciplinary Approach:
    - o Sustainability is a multidimensional concept that touches upon science, economics, social studies, and even the arts. By taking an interdisciplinary approach, educators can weave sustainability into existing subjects.
    - o Example: In science classes, students can learn about the water cycle, and then explore water conservation methods in agriculture. In economics, the focus could shift to the cost-benefit analysis of adopting sustainable practices.
- Problem-Based Learning:
    - o Engaging students in real-world problems related to sustainability can provide an in-depth understanding of the issues at hand. Problem-based learning encourages critical thinking and problem-solving skills.
    - o Example: Students can be asked to develop a sustainability plan for a local business or to come up

with innovative solutions to waste management issues in their communities.

- Community Engagement:
  - o Partnerships with local organizations, guest lectures, and field trips can provide students with practical experience and community perspective.
  - o Example: Students can participate in local clean-up drives, visit recycling centers, or even assist in community gardens to learn about organic farming practices.
- Project-Based Assessments:
  - o Instead of traditional exams, students can be evaluated based on sustainability projects they undertake. This method gives them a platform to apply what they have learned.
  - o Example: Creating a composting system for the school, implementing a school-wide recycling program, or developing educational materials that raise awareness about sustainability.
- Use of Technology:
  - o Online platforms and digital tools can be leveraged to make sustainability education more engaging and effective.
  - o Example: Virtual reality experiences can simulate the impacts of climate change, and interactive apps can teach students how to calculate their carbon footprint.

- Teacher Training:
  - o It's imperative for educators to be well-versed in sustainability topics so that they can effectively teach them. Professional development workshops focusing on sustainability can be highly beneficial.

- Example: Workshops can cover subjects like integrating sustainability into lesson plans, understanding climate change, and exploring renewable energy options.
- Parental Involvement:
  - Parents play a crucial role in reinforcing the values of sustainability at home. Schools can organize parent-teacher meetings specifically focused on sustainability education.
  - Example: Parents can be encouraged to participate in sustainability-related school events, or even volunteer to lead extracurricular sustainability clubs.
- Continuous Improvement:
  - The curriculum should be regularly updated to include new findings, technologies, and methods in the field of sustainability.
  - Example: Annual curriculum reviews can help educators adapt and update the sustainability components as needed.

Incorporating sustainability into the school curriculum is a long-term investment in the future, teaching the next generation to live and work in a way that is harmonious with the environment. It fosters a culture of mindfulness, making sustainability a part of the educational ethos rather than an add-on subject.

Community Involvement in Teaching Sustainable Agriculture

The involvement of the community in teaching sustainable agriculture has enormous potential to benefit both educational institutions and local populations. It provides a multifaceted learning experience that extends beyond the classroom and into real-world applications. Here's a

breakdown of how community involvement can be effectively
integrated into the educational framework for sustainable agriculture:

- Partnerships with Local Farms:
    - Establishing partnerships with local farms can offer
      students hands-on experience in sustainable agricultural
      practices.
    - Example: A semester-long partnership could involve
      students visiting a local organic farm multiple times to
      observe and participate in planting, harvesting, and soil
      management.
- Collaboration with Local Governments:
    - Local governmental bodies often have various
      sustainability initiatives, and schools can tap into these
      resources to enhance their curriculum.
    - Example: Collaboration could take the form of
      workshops on sustainable practices, guest lectures by
      local policymakers, or even community-driven research
      projects.
- Involvement of Community Organizations:
    - Various community organizations focusing on
      sustainability can be valuable resources for schools.
    - Example: Environmental NGOs can offer learning
      modules, field trips, and even volunteer opportunities
      for students.
- Student-Led Initiatives:
    - Encourage students to lead initiatives that engage the
      community in sustainable agriculture. This gives them a
      sense of ownership and fosters leadership skills.
    - Example: Students could organize a community garden,
      involving residents in growing local, sustainable
      produce. The produce could then be sold at a

community farmers' market, with proceeds going to a local charity.

- Service-Learning Projects:
    - o Incorporate service-learning projects that directly benefit the community while educating students about sustainability.
    - o Example: Students could develop a water conservation system to be implemented in a local park, educating community members about water scarcity in the process.

- Community Workshops and Seminars:
    - o Schools can host workshops and seminars that are open to the community, focusing on various aspects of sustainable agriculture.
    - o Example: Topics could range from composting and waste management to the economic benefits of sustainable farming practices.
- Parental Engagement:
    - o Parents can play a vital role in reinforcing sustainable practices taught at school.
    - o Example: Schools could organize "Sustainability Saturdays" where parents, students, and teachers come together to engage in sustainability projects.
- Media Outreach:
    - o Utilize local media outlets to spread awareness about the importance of sustainable agriculture and to highlight community involvement.
    - o Example: A school project that successfully implemented a composting system could be featured in the local newspaper, encouraging other institutions to adopt similar practices.
- Feedback Mechanisms:

- Continual assessment and feedback from the community ensure that the initiatives are meeting their goals and are beneficial for all parties involved.
- Example: Surveys could be distributed to community members who participated in a school-led sustainability event to gauge its impact and areas for improvement.
- Continual Learning and Adaptation:
  - As community needs and sustainability practices evolve, so should the community involvement strategies.
  - Example: An annual review involving all stakeholders can help adapt the strategy to changing circumstances and incorporate new best practices.

By deeply embedding community involvement into the framework for teaching sustainable agriculture, schools can offer a more enriching, practical, and impactful educational experience. Such an integrated approach fosters a communal sense of responsibility and stewardship for the land and its resources.

Case Studies: Successful Teaching Methods in Sustainable Agriculture

The Edible Schoolyard Project – Berkeley, California:

- Overview: This program transforms an unused schoolyard into a thriving garden and interactive classroom.
- Methodology: The curriculum integrates gardening into subjects like science, math, and history.
- Outcome: Students not only learn about sustainable agriculture but also about nutrition, which has shown to impact their lifestyle choices positively.

- Why it's Successful: The project makes sustainable agriculture relatable and applicable, instilling values of environmental stewardship at a young age.

Stone Barns Center for Food & Agriculture – Pocantico Hills, New York:

- Overview: The center runs an extensive youth education program alongside its agricultural practices.
- Methodology: School trips and workshops offer students a chance to engage with sustainable agriculture practices directly.
- Outcome: The students leave with a better understanding of where their food comes from and why sustainability matters.
- Why it's Successful: Real-world, hands-on experience drives home the lessons far more effectively than theoretical learning.

Farm to School Program – Multiple Locations, USA:

- Overview: The program connects schools with local farms to provide fresh produce for school meals.
- Methodology: Beyond supplying food, the program includes educational aspects like farm visits and gardening projects.
- Outcome: Schools report increased meal participation and greater student engagement with educational content.
- Why it's Successful: The approach turns every meal into a learning opportunity and engages other senses in the educational process.

Sustainable Agriculture Course – University of Vermont:

- Overview: This university course employs various teaching methods, from traditional lectures to on-field practice.
- Methodology: The course requires students to work on a real sustainable farm as part of their grading.
- Outcome: Alumni often pursue careers in sustainable agriculture, armed with practical knowledge and experience.

- Why it's Successful: The course combines theoretical and practical learning for a comprehensive education.

Green Bronx Machine – Bronx, New York:

- Overview: Founded by a teacher, this non-profit builds edible classrooms in underserved communities.
- Methodology: Students grow vegetables right in their classroom, integrating agriculture into subjects like science and mathematics.
- Outcome: The project has successfully improved school attendance, academic performance, and community engagement.
- Why it's Successful: The program directly addresses multiple challenges, from education to food security, making it a multifaceted solution.

These case studies reveal diverse and innovative ways to incorporate sustainable agriculture into education. While the settings and methods vary, all share the core principle of making sustainability an engaging, hands-on subject that students can actively participate in. These real-world examples illustrate the immense potential and success of interactive, experiential learning in fostering a new generation of environmentally responsible citizens.

Conclusion

As we conclude this comprehensive guide on sustainable agriculture, it is crucial to revisit the key topics that have been discussed. We began by laying the foundation with an understanding of what sustainable agriculture means and why it is essential in today's world. We delved into soil health, emphasizing its role as the cornerstone of any sustainable system. Various methods of soil management and conservation were presented, offering alternatives to traditional, often harmful practices.

We also explored water management, highlighting efficient irrigation systems and the responsible use of this vital resource. The chapters on pest and livestock management offered insights into balancing ecological systems while maintaining productivity.

Technological advancements were not overlooked; we examined how technology can enhance sustainability, yet also presented the challenges it poses. The discourse then moved to the economics of sustainable farming, discussing the feasibility and market trends that could make sustainable practices more accessible and profitable. Finally, we tackled the critical role of education in propagating these practices, showcasing successful teaching methods and community involvement.

Understanding these diverse topics collectively underlines the multi-faceted nature of sustainable agriculture. It's not just a method but a mindset that requires comprehensive planning, community involvement, and multi-disciplinary knowledge. The aim is to foster a future where sustainability is not an option but a necessity.

The Future of Sustainable Agriculture

As we stand at the crossroads of environmental degradation and innovative solutions, the future of sustainable agriculture holds both challenges and opportunities. The increasing global population and climate change will continue to put pressure on our food systems. However, the growing awareness and adoption of sustainable practices offer a ray of hope. A shift toward more localized, closed-loop systems is already underway, backed by technological advancements that allow for better data collection, monitoring, and automation.

Consumer preferences are leaning towards sustainably sourced and produced foods, thereby influencing market dynamics. This growing demand offers farmers the opportunity to benefit economically while also fulfilling their environmental responsibilities.

Government policies, research, and development are progressively aligned to support this paradigm shift. Initiatives like subsidies for sustainable practices and educational programs are making it easier for farmers to make the transition. The next frontier involves integrating advanced technologies like AI, blockchain, and IoT into sustainable agriculture, thereby optimizing resource use and minimizing waste.

The future is a canvas where technological innovation and traditional wisdom can come together to paint a picture of a sustainable, equitable, and resilient global food system. The journey is complex but necessary, and every step taken is a step toward a more sustainable world.

www.ingramcontent.com/pod-product-compliance
Lightning Source LLC
Chambersburg PA
CBHW062355290526
45794CB00005B/2237